About Island Press

Island Press is the only nonprofit organization in the United States whose principal purpose is the publication of books on environmental issues and natural resource management. We provide solutions-oriented information to professionals, public officials, business and community leaders, and concerned citizens who are shaping responses to environmental problems.

In 2002, Island Press celebrates its eighteenth anniversary as the leading provider of timely and practical books that take a multidisciplinary approach to critical environmental concerns. Our growing list of titles reflects our commitment to bringing the best of an expanding body of literature to the environmental community throughout North America and the world.

Support for Island Press is provided by The Nathan Cummings Foundation, Geraldine R. Dodge Foundation, Doris Duke Charitable Foundation, Educational Foundation of America, The Charles Engelhard Foundation, The Ford Foundation, The George Gund Foundation, The Vira I. Heinz Endowment, The William and Flora Hewlett Foundation, Henry Luce Foundation, The John D. and Catherine T. MacArthur Foundation, The Andrew W. Mellon Foundation, The Moriah Fund, The Curtis and Edith Munson Foundation, National Fish and Wildlife Foundation, The New-Land Foundation, Oak Foundation, The Overbrook Foundation, The David and Lucile Packard Foundation, The Pew Charitable Trusts, The Rockefeller Foundation, The Winslow Foundation, and other generous donors.

The opinions expressed in this book are those of the author(s) and do not necessarily reflect the views of these foundations.

ENGINEERING
the FARM

ENGINEERING
the FARM

**Ethical and Social Aspects of
Agricultural Biotechnology**

**Edited by
Britt Bailey and Marc Lappé**

ISLAND PRESS
Washington • Covelo • London

Library of Congress Cataloging-in-Publication Data
  Engineering the farm: the social and ethical aspects of agricultural biotechnology / [edited by] Britt Bailey, Marc Lappé.
      p.   cm.
  Includes bibliographical references.
  ISBN 1-55963-946-6 (hardcover: alk. paper) — ISBN 1-55963-947-4 (pbk.: alk. paper)
  1. Genetic engineering—Social aspects. 2. Agricultural biotechnology—Social aspects. 3. Genetic engineering—Moral and ethical aspects. I. Bailey, Britt. II. Lappé, Marc.
  S494.5.G44+   2002002893

British Cataloguing-in-Publication Data available.

Printed on recycled, acid-free paper ✪

Manufactured in the United States of America
09 08 07 06 05 04 03 02   8 7 6 5 4 3 2 1

For Melissa, naturally!
—B.B.

To Jacqueline and our children's safe future.
—M.L.

Contents

Acknowledgments

Engineering the Farm would not be possible without the generous support of the Richard and Rhoda Goldman Fund. Thank you!

We are particularly grateful to those individuals who provided encouragement, time, and talents in carrying our ideas into publication form: Chris Desser, Amy Lyons, Jill and Rob Hunter, and the rest of the Pangaea crew, Yarrow Sprinkling, Candy Lodge, Heidi Marshall, Noah Chalfin, and Steve May. We deeply appreciate their support in making this possible. We are also grateful to the Tides Center for its continued support.

Warm thanks and a bow to the contributors who not only braved curvy Highway 1 so that we could spend time together discussing this project in detail, but also endured the many months of edits, notes, and more edits. Without their perseverance and patience, we could not have the final project in hand. Thank you Lori, Shelly, Paul, Brewster, and Carolyn. We also thank those who became contributors during the second phase of this project: Peter, Norm, David, and Frankie.

We owe a gigantic thanks to Heather Boyer at Island Press. Her careful and sophisticated review of the manuscript resulted in critical improvements.

Last, though definitely not least, thanks and praise to all the farmers!

Preface

Britt Bailey

The transformation of agriculture from the hybridized conventional food crops to types that are now genetically engineered literally happened overnight. Before the mid-1990s, virtually no acreage was planted with genetically altered plants. In 1996, there were suddenly 500,000 acres planted. By the end of 2001, there were nearly 100 million acres planted globally.

According to the environmental community, we were witnessing a complete revolution in farming. To the industry creating genetically modified crops, the scale and rate of conversion merely represented the switch of a "tool." Just as hybrid seeds, the tractor, and chemical inputs were considered tools, so too was the tweaking of seed genetics. The biotech industry had been hard at work for more than a decade developing plants that would permit technological advances in weed control and pest protection. First onto the market were herbicide-tolerant plants (plants made to withstand the company's weed-killing brew) and plants with built-in insecticides. For industry, the resulting products were barely worth regulatory hubbub, and certainly did not require the heat of worldwide debate they kindled. But to consumers and environmentalists, debate, battle, and counterrevolution were precisely what was demanded by the new technology.

In a sense, I was one of the foot soldiers in this counterrevolt. In 1998, for the second time in two years, I found myself in the Netherlands discussing agricultural biotechnology with colleagues from around the world. I took a seat for my afternoon panel discussion next to an agronomist from Canada. He leaned toward me and said, "There is nothing wrong with biotechnology per se, it has just become a flash-point for all that is wrong with agriculture."

I leaned back in my chair and, to my surprise, agreed. Maybe agricultural biotechnology *was* simply taking the heat for a quarter century of environmental insults and economic dislocations thought to be caused by monoculture-style cropping systems and the overuse and dependence on chemicals and synthetic fertilizers. Biotechnology could be easily seen as a perpetuation of an earlier tool, a style of farming that had begun with the introduction of the tractor. Maybe agricultural biotechnology was simply the next step in the Green Revolution's utilization of chemical-demanding hybrid seeds. Either way you looked at it, at the close of the twentieth century, farming was suddenly in a state of siege.

But I disagreed with the flashpoint conclusion. I pondered the short discussion with the agronomist. There was something wholly different about biotechnology. Beyond the perpetuation of existing intensive farming practices, albeit with a biological twist, agricultural biotechnology presented a technological development requiring social discourse and thought. The public, however, was being asked to greet this new technology with open arms. One could almost see a billboard with an attached spotlight shining down on the words, "Hey world, look at the twenty-first-century way we are growing our food! How do you like them engineered apples?"

Before coming to the Netherlands, I had held daily discussions about agriculture and biotechnology with Marc Lappé. We did not find ourselves talking about the scientific data per se, although we found new studies fascinating and scrutinized them thoroughly. Instead, we traced the histories of our earliest memories of gardening (neither of us are farmers) and our memories surrounding food. I heard stories about his Grandma Hench's meals and his Grandpa Phillip's fresh-baked bread. And I responded with colorful recollections of summers spent with my grandmother in the coal-mining region of West Virginia. She taught me how to pull carrots and collect eggs from the chickens, and encouraged my game of leaping into haystacks. These were some of our memories of food and farming, and they were making their way into our attempts to understand the fuel of the arguments surrounding genetic engineering.

I was fortunate to grow up in Memphis, a city surrounded by fields of soybeans, cotton, and rice. By the time I was in high school, however, the downtown cotton warehouses were being turned into luxury condominiums, though the names of the businesses that depicted the bustling history of the city were still faintly readable on the sides of the brick buildings. One of the main financial institutions in Memphis is still called Union Planters. The farmers in the area had witnessed every step of technological progress over the years. On one trip home, I spoke with a local farmer who said he could not stay competitive unless he adopted the use of biotechnology. Bioengineered seeds were responsible for driving down the prices of final products. The biotechnology industry believes that its technology makes life easier for the farmer as well as decreases the overall price of inputs. But is it simply pushing farmers into a technology in order to stay competitive?

While farmers were trying to stay solvent, the public outcry surrounding bioengineered foods was getting louder. Marc and I believed that many of the reasons for public dismay and distress concerning agricultural biotechnology could be tracked to romanticized and likely more ecologically sustainable ideals about our food and farming. It was obvious to us that Europeans were in the throes of a new food scare, notably mad cow disease and dioxin-contaminated soda. Yet underneath their anxiety, Europeans were leading the advance for a cultural return to products that had not traveled halfway around the globe before finding their place on a Sainsbury shelf. In Europe, increasing numbers of people desire meat from animals that have not been raised under horrendous conditions requiring the widespread use of antibiotics or hormones. There has been an increasing demand for foods grown organically, or biodynamically. Being able to trace food on a supermarket shelf to its original soil is ever important to Europeans.

In the United States, the public, or consumers as they are now called, also appeared upset by the introduction of genetically engineered foods, but for different reasons than Europeans. The concerns were much more culturally oriented. Americans did not like discovering overnight that their food had been genetically engineered, leaving genetic by-products infused into their morning cereal and evening garden burger. They focused on the lack of democratic discussion prior to the food transformation, and questioned the corporate monopolies forming with each engineered commercialization.

In time, challenges to the introduction of genetically engineered seeds stretched into concerns of public health and environmental safety. What were the health effects, if any, from eating a gene from the germlike *Bacillus thuringiensis*? What ecological changes could we expect from the extensive shift in farming? I believe that scientific questions will continue to be asked, answered, argued, and answered again. Much of the polarization inherent in the topic is being staged on the results of scientific studies. Does *Bt* affect the immune system? Are monarch butterflies harmed by *Bt*-containing pollen? Will traditional weedy relatives of our domesticated cultivars become superweeds or extinct? Or will the impacts of biotechnology be more subtle?

In addition, Marc and I continue to believe the uproar surrounding biotechnology is fraught with fundamental social, cultural, and ethical issues not entering enough public discussions. Collective deliberations are necessary when entire systems, particularly those involving food production, shift and change. Questions about how we can protect and support indigenous farmers during the genetic shift, whether we should buffer centers of biodiversity, and whether we should be introducing a biotech farming style at all are but a few of the ones left out of industry's marketing equation. It was the recognition of this collective void that moved us to develop a grant that would fund a discussion among activists, scholars, and lawyers. It is our belief that much of the political and scientific pandemonium has occurred because the societal issues were shoved under the industrial carpet, rarely seeing the public light of day.

The Richard and Rhoda Goldman Fund has supported and graciously encouraged our work. In September 2000, at our office in Gualala, California, we convened a meeting of a dozen colleagues with varied perspectives and backgrounds who could pinpoint the most important social issues associated with biotechnology. The result of that meeting and hours of discussion and writing is contained in this book.

We hope these essays will shed light on topics that have not made the magazines or been demarcated by thirty-second news clips showing yelling activists in butterfly costumes and gas masks. Protesters seeking the media spotlight, yet knowing that regular folks in dungarees and T-shirts talking about their convictions is hardly news, are taking to the streets in full costume holding cleverly worded signs. In the fast and furious world of news stories, the necessarily complex understanding of social issues often loses out when a life-size glowing orange monarch butterfly or Greenpeace's monstrous green Franken-Tony the Tiger provides a better visual cue.

Behind the costumes and signs, people motivated by deep philosophical concerns have taken to the streets. Could the biotechnology industry ever have believed its innocuous bit of sci-fi tweaking would touch the common nerve? Agricultural biotechnology challenges our notion of food and the meals we share, its nutrition and our health. For many, genetically engineered foods threaten to uproot our traditional images of family, religion, culture, and society. The surge of the bio-engineered conversion threatens to unseat and override ecologically based methods of farming. Fundamentally, it challenges our relationship with this planet.

This book is intended to help depict the broader, societal story. Ideally, it will help shed light on the underlying convictions many people have about their food and their need to protect the increasingly fragile environment. It will also illuminate the growing monopolistic concentration of food production.

We discovered in the process of compiling and writing this book that the ethical and social issues associated with the technology are vast, complex, and often transcend traditional ethical analysis. Indeed, although there are scholars, philosophers, scientists, and lawyers among the contributors, none claims to be a formal "ethicist" in name or training. For many of us, biotechnological advances trigger ill-defined emotional stirrings. This deeper impact has motivated many of us to read beyond our traditional areas of expertise, to read ecological studies, treatises on biodiversity, patent and labeling laws, and endless policy documents both national and global. For those of us who strove for a foundational understanding of the challenges associated with agricultural biotechnology, nothing short of a deep search into the related annals of philosophy, law, agroecology, and anthropology proved sufficient.

We found the debate surrounding the infusion of new genes into crop seeds and food to be riddled with paradox.

- While some farmers are sold genetically engineered seeds in the name of lessening their costs, others are being bankrupted by the technology.
- Biotech foods are being promoted as a means to feed the burgeoning population, yet acres of genetically engineered seed seem to be yielding fewer bushels than before.
- Bioengineers point to the patenting system as a prime motivator for their inventive energies, but patenting removes the ability of many people to feed themselves with native varieties and cultivars of plants.
- Industry claims biotechnology is better for the environment, while activists and some researchers remain skeptical about its hidden ecological consequences.

- Agronomists point to a growing cornucopia of new seeds, but those trained in biodiversity remain undecided as to whether plant biotechnology threatens to reduce organism or genetic variation, or represents a tool that will rescue threatened landraces.
- Biotechnology companies claim their technology is more precise than conventional breeding and therefore should prove less threatening to public health. Some consumers and medical researchers believe new genes introduced to foods present potential novel allergens and health risks.

While the public has consistently stated its desire to have foods containing genetically engineered by-products labeled under a banner of a right to know, industry has exponentially increased production. Today, eating foods that *do not* contain genetically modified ingredients has become a virtual impossibility. While the public has demanded a right to choose foods free of engineered by-products, as contributor Paul Thompson points out in chapter 2, we have a stronger argument in a right to an exit from the pervasive engineered food system more generally. The only possible escape is the purchase of premium-priced organic foods. In chapter 4, Norman Ellstrand portrays how even organic crops may be tainted by genes flowing from engineered varieties. Genetic modification may herald the loss of entire ecosystems filled with traditional weedy varieties and landraces of plants as they become increasingly contaminated with pollen from new genetically engineered varieties.

Agricultural biotechnology represents a technological progress to some and disaster to others. For example, as Lori Andrews points out in chapter 5, the 1980 decision to allow patenting of genetically engineered organisms opened the door to development and commercialization of corporately owned seeds. Yet, the patent decision is also thought to have contributed to the scale-up of the bioengineered revolution, complete with corporations perceived as monopolistic seed dictators. Biotechnology has concentrated the global seed supply into the hands of a few corporations. For many people uneasy about the complete acceptance of the mix between technology and our food, bioengineered seeds represent a further dislocation from Nature. Carolyn Raffensperger and David Barling eloquently provide their passionate voices on this subject. Peter Rosset contributes a detailed and impassioned essay regarding the ability of bioengineered crops to feed the world.

We recognize the difficulty inherent in reviewing all of the various issues associated with agricultural biotechnology. What we offer here are voices of those concerned about the scale and scope of our newest transformation in farming. In our efforts to present diverse viewpoints, we have solicited essays from those outside academia, such as Brewster Kneen, Peter Rosset, and Carolyn Raffensperger. And, we have essays from those within academia, notably Lori Andrews, Paul Thompson, Sheldon Krimsky, Norman Ellstrand, and David Barling. We have brought together this group of writers because of our desire to broaden the base for understanding our societal response to agricultural biotechnology, and to move the discussion beyond the science and politics into the realm of social and ethical discourse.

Introduction: GMOs, Luddites, and Concerned Citizens

Marc Lappé and Britt Bailey

In the early 1990s, the newest Green Revolution was heralded by a spate of genetically engineered crops created in the laboratories of major producers such as Monsanto, DuPont, Rhone-Poulenc, and Aventis CropSciences. In spite of its science fiction connotations, the technique of genetic engineering was in truth "borrowed from nature." The earliest forms of plant crops were modified by capitalizing on the existence of *Agrobacterium tumefasciens,* a plant bacterium that had the capacity to insert its own DNA into a plant cell. The key attribute of the bacterium was its ability to have its own DNA sufficiently well ensconced that the plant itself would be tricked into making more bacteria as well as its own vital foodstuffs.

It was a short step from this realization, known since the 1960s, to using the bacterial insertion system to carry new genes into plants that had been selected by agricultural engineers. Thereafter, scientists conspired to come up with ever more elegant methods of co-opting plant genomes to accept and process human-selected DNA that would confer attributes of commercial or societal interest. It is precisely this tension—between the social and commercial values of newly engineered crop plants—that informs much of the early debate on genetic engineering.

Some key questions surround genetic engineering: To what extent did the pioneers from major chemical companies have an obligation to meet social needs in agriculture compared to assuring the commercial success of their own chemicals? What societal concerns about equity, distribution, and fairness does any new commodity have to serve? And

1

when a new commodity displaces an entire industry, as may now be the case with engineered food crops such as soybeans, should the producers have some form of societal consent?

From the beginning, the proponents of intentionally gene-modified food crops resisted the use of phrases such as "GMOs," or genetically modified organisms, to describe their inventions. The term GMO has been challenged by some biotechnology advocates as pejorative, because it implies that previous food crops were somehow not modified. Of course, many such plants were originally selected precisely because of their different genetic characteristics. At least one, the Asian pear, was expressly bred to allow the genes from a virus to enter its genome in order to combat a risk of blight. With these understandings, we nonetheless use the term GMOs in this book to describe gene-inserted plants and animals.

Proponents cite the prospect of these newly modified crops to transform agriculture, increase productivity, and shift our dependency on pesticides as justifying arguments for rapid expansion. One might reasonably ask whether such expectations were realized in the first decade of agricultural biotechnology. To some extent, one can argue that the expectations of agricultural biotechnologists have been fulfilled, as significant percentages of major staples such as soybeans and corn become replaced by engineered varieties, and the use of pesticides in other modified crops, notably cotton, has begun to shrink. But many environmentalists and activists remain concerned that GMOs carry hidden risks and costs to society, and may only serve to sustain our long-term dependency on chemically intensive agricultural systems.

Others decry the absence of a democratic process to guide and direct the use of genetic technologies, and challenge if the dimensions and extent of such use is currently warranted from a societal point of view. Still others question fundamental ethical issues in the dissemination of GMOs: Should any derived benefits of GMDs first be found to outweigh their risk of transforming other organisms in their environment.

The thoughts and writing of some of the most outspoken and insightful analysts and critics of the new revolution in engineered food crops have been assembled in this book. For the reader interested in the historical roots of the anti-biotechnology movement, the essays by Marc Lappé and David Barling provide perspective. The broad ethical issues in deciding where and when to introduce genetically altered crops are explored by Sheldon Krimsky. A related perspective is offered by naturalist Brewster Kneen, who discusses how agricultural and ecological values can sometimes clash. Norman Ellstrand provides an important essay that explores the risks to the ecosystem from genetic drift. Lori Andrews, a professor of law, offers a perspective on the role of patenting in permitting the explosive development of GMOs. Britt Bailey explores the question of health risks from GMOs, while Peter Rosset provides a critical analysis of the claim that bioengineered foods will feed the world.

Paul Thompson, a philosopher, argues forcefully about the minimum ethical condition for permitting GMOs to be marketed as consumer products. He argues that the consumer must have an "exit" from a system increasingly dominated by a single type of commodity that may not, for a variety of reasons, be acceptable. Carolyn Raffensperger, known mainly for her work in popularizing the precautionary principle, explores how a person might construct ethical arguments about GMOs that in part recognize Thompson's position.

Among the arguments in this book is the claim that many of these new genetically modified plants, such as Roundup Ready™ soybeans and their by-products, swept into the agricultural sector before the public—and the scientific community—had an adequate opportunity to debate their utility and safety. The industry counterpoint, that sufficient field tests and regulatory submissions were made to satisfy those federal agencies assigned the task of vouchsafing new products and pesticides, has defined the present status quo. We acknowledge that the agencies charged with overseeing the evaluation, field tests, or pesticidal properties of GMOs—respectively, the Food and Drug Administration (FDA), the U.S. Department of Agriculture (USDA) (especially the so-called APHIS program, which oversees field testing of crops), and the Environmental Protection Agency (EPA)—have never found sufficient public health or environmental concerns to suspend introduction, to require labeling or other disclaimers, or to restrict the presence of gene-directed pesticides in field crops except in one instance where a pest-protected plant was deemed fit for animal fodder but not for human food. This case, known as the StarLink episode, later gave regulators pause as the first example of widespread contamination of non-GMO crops with an unapproved gene product.

The current situation is a uniquely American status quo, one that permits market forces both here and abroad to shape the success or failure of GMO food crops. In Europe and Japan, in particular, governments have been loath to take at face value the equivalence of

GMOs to their traditional counterparts. They have introduced import restrictions and labeling requirements. Other countries, notably Canada, Australia, and New Zealand, have been urged by the United States to "go slow" and have delayed their labeling decisions until 2002 or later. However noble the concerns or opposition, U.S. governmental forces have encouraged these countries to accept GMO-based American produce on the strength of our indigenous experience and the assurance that we have done our part in reviewing the safety and non-invasiveness of the newly introduced genetic information. In part because of the overwhelming predominance of what is perceived by activists as a uniquely American-driven, pro-biotechnology position emerging among conservative world governments, the editors of this book believed it would prove useful to provide an array of critiques that question the present system.

In preparing this book, we have carefully selected contributors from a broad spectrum of political and activist persuasions. Not everyone has a blanket opposition to GMOs, nor do all of the writers come from the same political position or environmental perspective. In presenting this full range of opinions about GMOs, we hope to provide a litmus test for the cogency of the activist community's arguments. At a minimum, we hope to demonstrate that further debate of the ethical implications of GMOs is worthwhile and valuable.

A key factor in our decision to write this book is that consumers have had little opportunity to understand the full spate of potential environmental and public health consequences of new genetically engineered crops before being faced with the flood of GMOs now found on grocery store shelves. We also recognize that the reasons for the gulf between the two (or more) sides contesting genetically engineered foods are complex. Some critics believe that regulatory agencies such as the FDA, EPA, and USDA gave agricultural biotechnology a premature approval based on company assertions of safety and the "equivalence" of their products to preexisting food crops, before those claims could be fully scrutinized. To the activist community, the promotion of the belief in the absolute equivalency of GMOs to traditional foods, generally recognized as being safe, was seen as a corporate decision to expedite marketing rather than one based on sound science. From the corporate point of view, the testing and scale-up of production was fully lawful and reasonable, and was nothing more than business as usual.

The Battle Lines

This book explores the points of contention that separate corporate and activist communities. Some of the resulting battle lines may have been a "Two Culture" (as originally described by C. P. Snow) problem, separating scientists and humanists as two groups with irreconcilably different worldviews. It is likely that the alleged gulf between corporate self-interest and public accountability was more one of perception than reality. Indeed, many of our colleagues acknowledge that the issues associated with genetically modified foods initially became so polarized because activists could not accept otherwise "normal" strategic business decisions that have always been part of the corporate modus operandi.

The debate was also one over the degree of scientific acceptability: How safe is "safe"? What is a "sufficient battery" of tests? How is the apparent similarity between old and new crops to be evaluated? Such questions demonstrate that much of the GMO debate is not simply a matter of "taste" or politics. In fact, proponents and opponents often differed radically over the application of science, fracturing the battle lines over scientific values and the interpretation of data.

These issues of scientific accountability intensified in the early days of the debate between 1996 and 1998, when the battleground was divided along lines that contested the basic science. Critics ask, "Where is all the data that bear on the question of whether or not GMOs are 'equivalent' to prior conventional crops? How can we be certain GMOs pose no novel allergic risks?" Is it now too late to label some (or all) foods as containing GMO products? Will nontarget species, other than "a few" monarch butterflies, be harmed by pesticide-protected plants? Some of these issues appear partially resolved by recent scientific studies: Pollen containing the products of a bacterial pesticide gene may not be sufficient to produce significant attrition of nontarget insects such as monarch butterflies—but the presence of similarly created, wind-borne pollen tubes may still be a threat. Other issues, such as the question of labeling, are clearly a matter of politics and ethics, and as Paul Thompson points out, may not require any scientific testing at all to establish their validity. And the first generation of large, farm-scale field trials of GMOs (conducted in Great Britain) may provide ecological insights even as the field tests themselves remain controversial.[1]

Unfortunately, the picture painted by corporate biotechnology interests in the late 1990s was that such testing and evaluation was simply unneeded once crops like Roundup Ready™ soybeans had been de-regulated. As indicated by brochures, advertisements, and public relations campaigns of 1997–1998 (reviewed in Bailey and Lappé, *Against the Grain,* Common Courage, 1998), the corporate position was that these issues of concern were nonproblems, fully resolved by in-house testing. Today these assurances have been tempered by recent developments that we will later explore in depth: the controversial spread of genetically altered germ plasm from pollen-generating crops such as corn into nontarget species; bona fide health concerns about allergenicity from genetically inserted bacterial proteins; and the growing religious and personal concerns about food acceptability.

These realities have led the contributors of this book to debate the broader questions of corporate control, rights of indigenous peoples, control of genetic patents, and unintended contamination of seed stocks with genetically modified material. They give voice to these diverse concerns by exploring some of the social, political, and moral impacts of agricultural biotechnology. The overarching theme of their conclusions is that the scope, scale, and size of the present venture in modifying plant crop genes is so vast and intensive that a thorough review of agricultural biotechnology must consider its societal, cultural, and ecological impact as well as its effect on individual consumers. This impact includes the consequences of biotechnology's growing potential to dislocate political, ecological, agricultural, and nutritional systems.

Political Realities

The contributors also raise ethical concerns about the legitimacy of the public's right to know, which includes "rights" to choice and labeling; to having assurances of absolute or near absolute public and ecological safety; and to being granted access to corporate records and testing to assure accountability and transparency. Unfortunately, these issues were not forcefully addressed until after many companies had created a new kind of agricultural *realpolitik* in which a high percentage of staple crops such as corn, sorghum, and soybeans were already committed to genetically engineered crops. By the end of the year 2000, 75 percent of the soybean crop, 30 percent of the corn, and 10–15 percent of sorghum had been genetically engineered; in 2001, some 25–30 million hectacres were planted worldwide with these and related crops.

To some observers, this political reality makes any meaningful reconsideration of the technology largely academic. A decade after the first genetic food commercialization, this book challenges this fait accompli premise by asking fundamental questions and returning to basic issues raised by the new genetic revolution in agriculture: What is the proper relationship between farming and society? Can anyone properly "own" or patent new plant varieties? Do consumers have the right to have an "exit" to the system of genetically engineered foods, irrespective of the "rationality" of their decision? When does control over a commodity become an unacceptable monopoly? Should patents continue to be offered on "new" varieties of food crops. How are the presumptive benefits of this new technology (e.g., the feeding of the world) to be assessed against its global risks? And, ultimately, to whom does the germ plasm of food crops belong?

Ethical Concerns

Ironically, in the early twenty-first century, evincing ethical concern for issues such as the ones mentioned above has become de rigueur for contemporary ag-biotechnology CEOs. In the face of concerted public opposition, some corporate executives have shifted their rhetoric, promising greater corporate accountability, transparency, and even a modest accommodation to consumer interests. In some sense this corporate sensitivity to moral issues is an affirmation of the legitimacy of earlier critics, but some of this corporate concern still appears more about public relations than true accountability. Activists, including some of the authors in this book, have dismissed the corporate pronouncements of public accessibility as doing little to move the debate on ethical issues forward. For many of us, this skepticism is informed by historic precedent where some leaders of major corporations discounted the legitimacy of public concerns about genetically novel foodstuffs, including fundamental questions of the right to know about food composition, safety testing, and true long-term biocompatibility with sustainable agricultural practices.

From the industry side, corporate spokespersons insist their statements do indeed voice a new ethical concern intended not merely to alleviate residual angst about biotechnology, but to offer greater public responsiveness. Time will tell. Whatever their ultimate positions, both camps readily admit that the debates around genetically modified organisms have intensified. In dozens of incidents, "eco-terrorists" have destroyed crops, seed laboratories, and even small forests in an attempt to thwart the expansion of genetic technology, acts that many of the activists writing in this volume have publicly decried. Some corporations or their employees may be responsible for equally egregious acts. According to recent reports, some biotech purveyors have allegedly encouraged the wanton spread of unapproved GMO crops into newly emergent countries such as Mexico, where the nonnative engineered pollen threatens to contaminate indigenous germ plasm.[2] Other companies have allegedly hired special protective services to monitor fields and litigiously pursued those (including ourselves) who have allegedly sullied their products in print.

Luddites and Critics

Critics of the critics assert that anyone who claims genetically modified foods are unsafe or environmentally dangerous is a Luddite looking for a new cause célèbre, because the scientific data to support harm or hazard is not there. Activists, in turn, say that if the data is not there, it is because the premarket testing that should have been done was not done, and that regulatory agencies such as the EPA concede enough residual risk remains to warrant continued vigilance.

Many of the contributors in this book would probably rankle at the "activist" label, or the suggestion that they are merely opposed to technology for anti-technology's sake. Indeed, some of the writers would argue that more science and technology are just what is needed if we are to resolve questions of safety, persistence, and spread of genetically modified food crops. Unless and until such tests are complete, many consumer groups have been arguing for labeling as a concession to consumer rights. Paul Thompson goes further by observing that allowing consumers "exit" is an even more fundamental issue that cannot be reduced to a simple issue of "freedom of choice."

To do so, Thompson advocates labeling of GMOs as a necessary but not sufficient solution. But this answer is complicated in practice. If genetically modified foods are to be properly identified, they will require highly specific tests to measure potential contaminants. An especially high standard of scientific accuracy is needed to establish the labeling standards for GMOs demanded by activists. For instance, the recent decision by the European Union to require labeling of foodstuffs containing GMOs of 1 percent or greater will require highly specialized testing to avoid ambiguous data and possible mislabeling.[3] In practice, the accurate detection of the true number of copies of new "transgenes" is a daunting task, especially for small sample sizes.[4] Perhaps of even greater concern is the recent revelation that totally unexpected foreign "gene fragments" can be found after engineering soybeans and perhaps other crops such as cotton.

But should these technical concerns decide the issue of separating "contaminated" from "GMO-free" crops? Some would insist that public disclosure of any kind is preferable to silence. Others would say disclosure is reasonable only if it is scientifically verifiable and accurate. Some of these concerns are obviated by the newest rulings, such as those of the European Union on labeling that allow a 1 percent leeway in contaminant levels.

With or without labeling, a recent Harris survey establishes that a great majority of persons (up to 80 percent) are aware they probably have ingested GMO foods or by-products.[5] Whether or not consumers are upset about this likelihood remains largely a matter of social forces that shape and influence public opinion. Whether consumers should be concerned about food safety is an ethical, not a scientific, question. Whether they should be alarmed, concerned, or merely curious about the "slight" alterations in chemical composition of their food is both a scientific and a social question of tolerance for often ambiguous data about low-level risk.

Too often, the answer to such questions is left to the propagandists on both sides of the argument. From the industry, we hear assertions of the "substantial equivalency" of engineered and nonengineered varieties. From those opposing the technology, we learn of subtle differences that *could* make food dangerous and "unnatural." Behind the rhetorical issues are bona fide health concerns about allergenicity, where the very newness of GMOs creates a risk of activating a naïve and inexperienced immune system.

The opponents' concerns that some engineered foods may be allergenic—notably, engineered corn containing Cry9C because of its structural novelty and persistence in the intestinal tract—was iterated by the National Research Council and has recently been echoed by a prestigious panel assembled by the EPA. Still other concerns, notably those about novel composition, unforeseen ecological impacts (especially on the subsoil microbial communities), and potential nonequivalence for certain key biological ingredients (e.g., phytoestrogens in soybeans) all have growing scientific credibility. The key question is not about the nature and extent of the risk posed by genetically modified foods (which is currently unresolved), but what should be done in the interim in the present environment of uncertainty.

The existence of consumer concern or anxiety alone does not create a sufficient basis for removing incompletely studied products from grocery shelves. Similarly, a label that discloses only the genetic status of the food does not convey adequate consumer information about the relative safety of that food, especially if testing of GMOs has been inadequate in pointing to meaningful conclusions about the presence or absence of health risks. Nonetheless, the governments of Japan and the countries of the European Union appear satisfied with simple disclosure as a solution to consumer concerns. The U.S. government remains steadfastly opposed to any mandatory labeling standard. As some of the authors in this book point out, deciding whether or not to label is more than a scientific issue: it is an ethical and political one. The following chapters explore a full gamut of such issues in an effort to provide the range of often divergent viewpoints that make up the dissent to biotechnology—or at least to its current form of implementation.

Chapter One

Ethical Issues Involving the Production, Planting, and Distribution of Genetically Modified Crops

Sheldon Krimsky

The discovery of plasmid-mediated gene transfer in 1973 afforded science a revolutionary technique for rearranging and modifying the genetic structure of biological species.[1] Other techniques for transferring genes followed, including the use of viruses, DNA projectiles, and microinjections. Thus far, there appear to be no natural or species barriers limiting the transfer of genetic material across organisms of different phyla and even kingdoms that cannot be overcome by the set of processes known as recombinant DNA or gene transplantation technology.

Agriculture was one of the first industrial sectors to have invested heavily in the new field of biotechnology. By the early 1990s a massive experiment in agricultural biotechnology was underway in which a new generation of crops containing genes from sources outside the plant species was introduced into food production in many parts of the world. The genes transferred include some that express new proteins, some that mark specific parts of the genome (marker genes), some that regulate gene expression (e.g., promoter sequences), and finally some that provide identifying clues that the gene transfer has been accomplished (e.g., antibiotic resistance genes).

We are at the early stages of this global agricultural experiment.

Scores of new food products with altered phenotypes are slowly moving from genetics laboratories into commercial products. By the year 2000, approximately one-fifth of the U.S. corn acreage, one-half of the soybean acreage, and three-quarters of the cotton acreage, comprising nearly 30 million hectares, was planted with crops genetically modified for resistance to insects and tolerance to herbicides.[2]

This global agricultural experiment in biotechnology has been met with controversy in Europe, parts of Asia and South America, Australia, Canada, and New Zealand. This essay explores the ethical and value components of the controversies that have erupted in the wake of the first introductions of genetically modified (GM) crops since the early 1990s. These controversies have affected international trade agreements and have divided environmentalists.

Among the issues that have spurred some of the most highly contested debates are the following: (1) the ecological effects of releasing GM seeds into the environment; (2) the impact of GM crops on global seed markets; (3) farmer and consumer preferences in the adoption of GM products; (4) the role of risk assessment in evaluating the safety of transgenic seeds; and, (5) the impact of the global use of genetically engineered crops on biodiversity.

In recent years science policy analysts considered it possible and desirable to separate the scientific from the ethical issues in science and policy studies; I find this cannot be easily accomplished for the controversies involving biotechnology. The normative and the empirical parts of the biotechnology disputes are tightly interconnected. Sometimes the empirical issues provide false cover for the normative questions. Other times the value conflicts are based on disputed scientific claims. Many of the ethical issues involved in the political debates over GM foods/crops are not sui generis but depend on the resolution of empirical questions.

The one contested issue involving GM crops that comes closest to resting on purely ethical considerations is whether it is morally permissible (irrespective of consequences) to alter plants by genetic engineering technology. Human rights and animal advocacy groups have proclaimed the genomes of humans and animals as inviolate for human genetic manipulation. Their moral justification rests on "natural law" (e.g., species nature or the sacredness of human germ cells) or consequentialist arguments such as the uncertainties that may result from tampering with nature.

Others have appealed to a secular repugnance for bioengineered plants.[3] Those who claim that applying gene transplantation processes to the germ plasm of crops violates the natural order might be hard pressed to apply the same standards to the other human interventions during the ten thousand years of plant domestication. Are there morally distinguishable issues that make the current techniques of gene modification a transgression against the natural order and the earlier ones not? How are human-selected gene sequences different from those made by hybridization, chemically or radioactively induced mutations, cell fusion, or synthetic foods? An issue that bears directly on whether GM crops/foods should receive special ethical status is the uniqueness or lack thereof of plant germ plasm modified by gene engineering techniques.

Issue 1: Are GM crops/foods unique?

The question of the uniqueness of genetically modified organisms (GMOs) may be divided into two parts. Are compositions of GMOs unique? That is, by applying recombinant DNA processes, can a product be made that would not otherwise be found in nature or that could not otherwise be constructed by other techniques, such as conventional plant breeding? The second part of the uniqueness issue pertains to whether the risks of GMOs to human health or to the environment are unique. Will the introduction of GMOs to the biosphere produce novel hazards?

Three reports of panels convened by the National Academy of Sciences (NAS) concluded that the use of genetic engineering techniques to produce crops do not result in any unique risks in comparison to techniques of conventional plant breeding. The first report issued by the NAS Committee on the Introduction of Genetically Engineered Organisms into the Environment was published in 1987.[4] A second, longer study was released in 1989, also by a committee of the NAS.[5] Finally, a third study, released in 2000, was titled *Genetically Modified Pest-Protected Plants: Science and Regulation.*[6]

The 1989 NAS report stated that "no conceptual distinction exists between genetic modification of plants and microorganisms and classical methods or by molecular methods that modify DNA and transfer genes." It also stated, "crops modified by molecular methods in the foreseeable future pose no risks significantly different from those that have been accepted for decades in conventional breeding."[7] The conclusion was reaffirmed in the third NAS report, which highlighted two points: (1) There is no evidence that unique hazards exist in either the use of rDNA techniques or in the movement of genes between unrelated species; and, (2) the risks associated with the introduction of rDNA-engineered organisms are "the same in kind as those associated with the introduction of unmodified organisms and organisms modified by other methods."[8]

Although still a contested issue within scientific circles, the claim that there are no unique risks to rDNA techniques has been a key factor in shaping regulatory policy. Genetically engineered crops are regulated by one of three agencies (the Food and Drug Administration, the U.S. Department of Agriculture, and the Environmental Protection Agency) by and large in the same manner that conventional crops are regulated. There is only minimal pre-testing of GM crops. Because there is a presumption the transgenic food products are safe, a strong burden is placed on those who question the safety of the GM food to demonstrate the risks.

What is the basis upon which points 1 and 2 are accepted? Is there scientific evidence or is it based on a trans-scientific argument? The question of unique hazards breaks down into two parts: (1) Are there hazards? and (2) Are they unique? The issue of whether there are hazards in using rDNA techniques has been resolved in the affirmative (e.g., the Brazil nut allergen transferred to a soybean).[9] Are the hazards from the rDNA process unique? A reasonable interpretation of the meaning of *unique* can be framed by asking if a hazard can arise from conventional methods of genetics. Has anyone tried to produce the same results by conventional methods? If it hasn't been tried (successfully or not), how can one know that it is or is not a unique hazard? Besides the use of rDNA techniques, how else would the Brazil nut gene enter the soybean? Is that gene found naturally in soybeans?

In another interpretation, *unique hazards* might refer to a general class of hazards and not any particular one. Under this interpretation it is not inconsistent to state that rDNA technology is the only known way to transfer the Brazil nut allergen to the soybean but it is not the only technique that can transfer allergens from one crop to another, and thus does not introduce unique risks.

Until the term *unique hazards* is clarified and the empirical questions pertaining to non-rDNA methods for transferring allergens are answered, the query "Are GM crops/foods unique from the standpoint of hazards?" remains unresolved. If GM products were unique compositions of matter resulting in unique hazards, there could be ethical reasons to treat those products differently than conventional crops/foods.

This example illustrates the interrelatedness of the normative and empirical dimensions of the problem.

Issue 2: Does society have a right to hold transgenic crops to a higher standard of oversight than conventionally bred crops?

Putting aside whether or not there are unique hazards, it is clear that many public interest groups and the majority of the public in a number of countries believe transgenic crops should be held to a higher standard than conventionally bred crops. To say that rDNA techniques produce unique hazards does not imply that there are no hazards associated with conventional breeding (e.g., hybridization or cross-pollination). The use of rDNA technology in food production may deserve more oversight because it is newer and less rigorously tested than are other methods of crop modification that have been in place for much longer periods. Moreover, even if one were to assert that rDNA techniques do not produce unique hazards, one might still wish to give greater primacy to the hazards of genetic technologies over those of conventional breeding because of the very novelty of the risk potential entailed by the specific gene transfer. Society makes all sorts of risk selection choices based on collective values and perceived risks.

Do regulatory agencies bear a responsibility to respond to public demand for more oversight over GM products? The jurisdiction of regulatory agencies is established through legislative mandate. Within the boundaries of their jurisdiction, agencies make choices. Health agencies decide what goes on food labels. European and U.S. labeling standards are distinctively different, although both respond to health promotion. European labels focus more attention on chemical residues whereas U.S. labeling has a strong emphasis on nutritional content. The priorities agencies set often respond to public perceptions of risk.

Agency personnel and others who comprise the "community of experts" may differ with the public in setting public health priorities. But in democratic societies, even a consensus of elites must defer to the voices of popular opinion. Examples where public concerns influenced agency decisions include the safety standards for nuclear power plants and the risks of toxic waste sites. In both cases the public's concerns about safety exceeded and predated those of regulatory agencies. Eventually, the government's policies became more in step with public concerns.

In the case of GM crops/foods, public risk concerns in the United States and Europe exceeded those of their respective regulatory bodies.

This was clearly illustrated when the USDA withdrew its initial proposal for new federal organic labeling standards that would have included GM products under the organic label. The GM crops/food policies developed by U.S. regulatory bodies were heavily influenced by large biotechnology corporations.[10] When there are sharp differences between agency and public views over risk, governments have resources at their disposal to influence public opinion. However, when public skepticism persists, as it has with respect to genetically modified crops/food, then the representative bodies and their executive branches have an ethical responsibility to recalibrate their priorities in order to meet the democratic mandate.

Issue 3: Do people have a right to exclude themselves from the experiment?
Once again, setting aside the question of whether there are hazards or unique hazards associated with GM crops/foods, do people have a right to exclude themselves from this experiment with the global food supply? Suppose that a GM product meets regulatory standards. Are there any ethical grounds for giving consumers a choice over whether they consume the GM product? In many areas where new drugs, new foods, and new technologies are introduced, consumers have had a choice to be first users, last users, or nonusers. This has been the case with the introduction of the synthetic fat (Olestra) used as an oil substitute in chips, as well as sugar substitutes, which have been approved by the FDA. The premise behind the proposal to label GM foods is based on the idea of consumer sovereignty, namely, that people have a fundamental right to know what they are eating, how it was produced, and whether there are any uncertainties about its health effects.

Countries that have adopted labeling include Japan, South Korea, the European Union, Australia, and New Zealand. We label foods for many reasons other than the nutritional content. From public opinion surveys, a majority of Americans seem to support labeling.[11-13] On what ethical grounds is a labeling policy dismissed? Is there a conflict between the FDA's statutory mandate for labeling and the conditions of production for GM foods? Is the FDA forced by its statutes to reject labeling of GM foods, or has the agency interpreted the law in a way that favors industry's interests?

According to the FDA, a label must be materially relevant to the safety or nutritional value of a food product. In its 1992 policy on bioengineered foods, the FDA stated that "[it] has no basis for concluding that bioengineered foods differ from other foods in any meaningful or uniform way, or that as a class, foods developed by the new techniques present any different or greater safety concern than foods developed by traditional plant breeding."[14] The agency has historically interpreted the term "materially relevant" to mean "information about the attributes of the food itself," and has required labeling where the absence of information poses health risks or misleads the consumer because of other information provided by the manufacturer.

In addition, the law states that the label cannot make or imply false health claims about a food product. On these grounds the FDA opposed mandatory labeling on milk produced with the aid of synthetic bovine somatotropin (rBST), commonly known as synthetic bovine growth hormone, or any other food developed using bioengineering, with some minor exceptions for cases where there have been material changes in nutritional quality or the introduction of an unexpected allergen. The FDA also opposed voluntary labeling unless it contains language stating there is no health or nutritional difference between the bioengineered and the nonbioengineered food product.

The FDA is not opposed to labeling irradiated food on grounds of "materiality." Although it has consistently held that irradiated food is not nutritionally inferior to its natural source, with regard to irradiation the FDA concluded that it "could cause changes in the organoeptic properties of the finished food and that without special labeling, consumers might assume that such foods were unprocessed."[15] As of September 2000, the FDA reported that it had no data or other information that would support a regulatory decision that food or its ingredients produced using bioengineering meets its statutory criteria for mandatory labeling. But considering the strength of public opinion, the FDA acknowledged that "providing more information to consumers about bioengineered foods would be useful."[16] To resolve the conflict between the public's desire and the agency's labeling requirement, the FDA proposed a guidance document to "assist food manufacturers who wish to voluntarily label their foods as being made with or without the use of bioengineered ingredients."[17]

Without mandatory labeling of GM foods, consumers do not have a *right* to extricate themselves from the experiment on the food supply. They can, however, make an effort to purchase organically produced food, which, at least currently, is certified to be 99 percent GM free. From an ethical standpoint, consumers in the United States are not afforded a *right* to GM-free food. Only those consumers with access to organic foods have an *opportunity* to purchase GM-free products.

Issue 4: What ethical responsibility has society to address genetic pollution?
A farmer planting non-GM seeds may find that some of his yield consists of GM crops either from seeds or pollen that was deposited from a neighboring farm. The trespass of unwanted GM germ plasm to a non-GM farm is referred to as genetic pollution.

While the issue of genetic pollution is new to legal systems throughout the world, there is precedent for intraparty compensations from environmental externalities. Pollution by genetically modified pollen may constitute a taking—a legal theory currently being tested in the courts. Thus, a company that contaminates a water supply, which adversely affects another company's production, may be subject to liability even if the polluter is complying with the law.

The National Farmers Union in Canada is supporting action by the federal government to make agricultural biotechnology companies financially responsible for contamination of organic and traditional crops by GM-based agriculture.[18] In 1999 the British Broadcasting Company and Friends of the Earth employed a German laboratory to conduct DNA testing on various farms. The research showed that pollen from a GMO canola field ended up 2.8 miles away in a beehive.[19]

Currently, the only means through which consumers can be reasonably sure that their purchases of primary agricultural products have not been grown from GM seeds is if they buy organic produce. But that doesn't guarantee that transgenes have not contaminated the organic farms. Scientists have reported that at 50 meters from a small plot of genetically modified plants of oilseed rape containing a herbicide resistance gene, about one in ten thousand seeds produced by the surrounding nongenetically modified oilseed rape plants showed resistance to the herbicide.[20] It is uncommon for pollen to be transported more than a few kilometers, but it does occur during unusual weather conditions when the pollen is swept high enough in the air. Writing in the journal *Nature*, a team of scientists from the Scottish Crop Research Institute stated: "Our results show that significant quantities of pollen travel over large distances; this has implications for transgene recruitment by feral populations, provided pollen viability and competitiveness are unaffected by dispersal."[21]

Organically produced food is one of the fastest growing sectors in agriculture. Since public demand for organic produce has increased, the social institutions bear some responsibility to protect organic farmers from controllable externalities like genetic pollution. This type of protection may require the establishment of buffer zones separating farms or the use of sterile seeds for GM crops grown in proximity to non-GM farms. Once the case law develops, it is likely that a new set of agricultural norms will emerge that will provide guidance both for farmers who plant GM crops as well as for those who do not plant them. Liability claims will establish new risk-benefit ratios from transaction costs such as insurance for farmers planting GM crops. However, if farmers who do not plant GM crops are not protected from genetic pollution by the courts or by new legislation, their only option may be a self-initiated sequestration of organic producers to safe regions far enough away from GM farms.

Who bears the responsibility for this type of agro-genetic contamination? Organic farmers in North America can lose their certification if their crops show greater than 1 percent GMOs. In Germany and Japan the standards for certifying crops as organic require that they contain less than 0.1 percent of GMOs.

The USDA has established uniform standards for produce to be labeled organic. At one stage in this standard-setting process, the USDA proposed including GM products under the organic label if they were not grown with synthetic pesticides. However, the USDA retracted that provision in reaction to an overwhelming response from consumers that they wanted organic food to be GM free.[22] Does this imply that the government has an obligation to insure that organic food can be grown without GM pollen contamination? Can farmers seek compensation if their crops have been affected by GM pollen? Is there a "statute of limitations" for farmers who seek indemnity for GM pollution? Who is responsible for testing the food supply to ensure that organic foods are GM free? Organic food organizations are asking for legislative relief against genetic pollution of their crops.

These issues are currently being debated among legislative bodies and contested in the courts but have not yet been resolved. From the standpoint of agricultural ethics, farmers should have a fundamental right to farm natural foods without contamination from industrial or other agricultural sources. If GM pollen and seed dispersal become the norm, then this right will become meaningless.

Issue 5:What ethical principles guide intellectual property ownership of seeds?
Seed manufacturers have devised a strategy that protects their intellectual property in germ plasm and can protect organic farmers from GM pollution. They have been able to produce plants whose pollen and seeds are sterile. Once the GM seed is planted, farmers will not be able to save seed for a second planting. At the same time, the sterile pollen of those plants will not be able to outcross with non-GM plants.

Public interest groups have termed the device "terminator technology." It has been argued that sterile seed technology reduces the options for farmers and gives seed companies too much control over what a farmer can plant. Alternatively, the ag-biotechnlogy industry argues that seeds (GM or otherwise) deserve as much intellectual property protection as any other potential product (e.g., software in the computer field).

The only thing holding back biotech companies from producing and distributing "terminator" seeds has been the intense negative publicity, since these seeds do not violate any international or domestic laws and are not inherently more ecologically hazardous than ordinary GM seeds. Paradoxically, terminator seeds may be environmentally safer than their nonsterile counterparts. This likelihood raises an ethical dilemma for some environmentalists who are faced with an issue of containment versus farmer rights to save and replant derivative seeds.

The terminator technology has been of interest to USDA, which announced in August 2000 that it would partner with Delta Pine Land Co. to pursue commercialization of a "technology protection system," their term for terminator technology. In their system three genes are added to a plant, which, if treated with antibiotics, will produce a toxin that renders the subsequent generation of plants sterile.

Do the seed manufacturers have a right to protect their seeds as intellectual property by genetically engineering germ plasm from being repropagated in a second-generation planting? Is there a higher moral duty that gives farmers the right to use the plant germ plasm that they purchased in any way that enhances their agro-ecosystem and maximizes their utility? Can this be done without violating provisions of the patent law, for example, if farmers do not develop and sell commercial varieties based on the original GM seed? On the other hand, do seed companies have a right to market sterile seeds? Farmers who do not want those seeds can buy elsewhere. For American farmers, second-generation high-yield hybrid seeds do not provide the same yield as the first year, largely because of genetic variation introduced in F_2 crosses. As a result, U.S. farmers have become accustomed to purchasing new seeds each year. This is not true for farmers in developing countries. Thus, the ethics of terminator seeds may depend in part on whether it is applied to Third World or First World agricultural systems.

Another consideration of terminator technology is whether it will eventually be applied to domestic animals (pets and livestock). In such instances, farmers and pet owners will lose breeding rights for animals they own. These applications may be more restrictive to First World farmers than sterile seeds.

Is terminator seed technology in the public interest? Is there an ethical distinction between (1) forcing farmers, through contract, not to use the GM seeds of harvested plants, and (2) the practice of developing sterile seeds (terminator technology) so that farmers are unable to use those seeds?

The autonomy of the farmer to produce safe and nutritious crops must be a high priority of our system of agricultural ethics. Autonomy implies expanding and not narrowing choices. Since so few American farmers save second-generation seeds, there is no loss of choice in introducing sterile seed technology. Many of the U.S. seed manufacturers require farmers to sign contracts that they will not save their GM seeds, sell them for research, or plant them in ways the seed companies do not approve. There is, however, a loss of autonomy to organic farmers if they are faced with uncontrolled genetic pollution of their crops. The ethical arguments look different for Third World farmers for whom seed ownership of any form is culturally unacceptable.

Issue 6: Wherein lies the responsibility to stop the treadmill of resistant organisms?

Scientists have discovered products and techniques that destroy microscopic bacteria, weedy plants, and troublesome insects. But these same products and techniques, if used often, can help nature select resistant strains of organisms that reintroduce the problem. Cases in point are herbicide-tolerant plants. Among the first commercialized products of agricultural biotechnology, herbicide-tolerant crops are attractive both to industry and to some farmers. For the industrial sector, a single broad-spectrum herbicide will centralize chemical inputs and create higher profit margins. Farmers who can use a single herbicide for all their crops will suffer fewer losses from rotational planting of two crops that have tolerances to different herbicides. By applying broad-spectrum herbicides, two ecological effects are likely to occur. First, weeds that are naturally resistant to the herbicides will grow more readily and proliferate without the competition of the nonresistant strains. Second, the genes that confer herbicide tolerance to the food crop will transfer to weeds, also creating resistant strains.

A second case centers on the proliferation of antibiotic-resistant genes (as markers) in GM crops. The antibiotic-resistant genes may get transferred to the stomach bacterial flora of humans or animals. This will exacerbate the population of bacteria, some pathogenic, with resistance to therapeutic antibiotics. Overuse of a good product such as antibiotics, whether in pharmaceuticals, in antibiotic soaps and cleansers, or in plants can result in a negative outcome. The proliferation of antibiotic-resistant strains of bacteria has become a formidable public problem.[23]

A third case is the introduction of insecticidal genes, such as the gene that synthesizes the toxin for *Bacillus thuringiensis* (Bt). Plants with the Bt gene expose insects to an insecticidal protein at every stage in the plant's development and throughout the entire growth cycle. This will eventually create an evolutionary pressure that favors resistant insects, unless some accommodation is made.

Under whose responsibility is it to minimize the creation of organisms resistant to control agents? The problems of resistance are exacerbated by a number of human activities, including the overuse of antibiotics and antibiotic markers. For example, in 1976 a citizens committee in the city of Cambridge, Massachusetts, drafted the first legislation in the United States regulating rDNA research. One of the provisions of the ordinance was that antibiotic-resistant markers used in the creation of rDNA organisms not compromise the value of clinical antibiotics. Currently, GM crops use antibiotic-resistant genes—some of which may confer resistance to widely used antibiotics.

Do we have an ethical obligation to minimize the proliferation of resistance genes for antibiotics, insecticides, and herbicides? Who bears the responsibility for protecting society against the spread of resistant organisms? If alternatives to antibiotic markers in the development of GM crops are available, should their use in plants be mandatory? Should agriculture be moving in a direction that creates more evolutionary pressure for creating organisms resistant to biocides, for example, by exposing target organisms over a long time span throughout the growing season to these biocides?

The U.S. EPA has linked the registration of varieties of Bt crops with resistance prevention strategies. The agency is aware that the loss of Bt effectiveness through growing insect resistance could mean the return to higher toxicity pesticides.

Issue 7: When is the introduction of GM foods to the Third World ethically and politically correct?

It is estimated that one-fifth of the world's population is undernourished or living under famine conditions. The author of a recent feature article in *Time* magazine (July 31, 2000) stated that Golden rice (rice modified with the addition of the gene for beta carotene, a building block for vitamin A) is "the first compelling example—of a genetically engineered crop that may benefit not just farmers who grow it, but also the consumers who eat it" (p. 41).[24] *Time* reported that Golden rice could help at least 1 million children who die annually from vitamin A deficiency and an additional 350,000 who succumb to blindness. Some 3 billion people depend on rice as a staple food, while 10 percent, or 300 million, are afflicted with some form of vitamin A deficiency. What are the ethical conditions that define this issue? What are the benefits of introducing Golden rice? Will the necessary conversion of beta carotene to vitamin A occur in malnourished infants? What are the risks? Is Golden rice a ruse product designed to win over the world's approval for GM crops? Is it an authentic humanitarian product? Under what conditions would we accept a vitamin-enhanced rice as humanitarian? New Zealand does not allow any enhancements to its food—no vitamin-enhanced milk or vitamins added to grains in cereals. Would this country be ethically remiss to ban the introduction of genetically engineered, vitamin A–enhanced rice? How would this differ from conventional vitamin enhancement?

One commentator noted that GM-crop risks, if there are any, are relatively insignificant to people who are starving or who have severe nutritional deficits. Should the ethics of GM foods be calibrated to the desperation of people? We use a similar ethical approach in drug development. People with life-threatening diseases take more risks with experimental drugs than healthy people would be permitted to take. How should decisions about exporting new strains of rice to desperate nations be made? Should there be an international ethics board (like our local IRBs that review clinical trials)? If one is not opposed to GM crops/foods in principle, then what are the conditions under which it is ethically acceptable to send GM rice to developing countries to prevent or reduce vitamin deficiencies?

The humanitarian impulse to prevent vitamin A deficiency in children living in impoverished regions of the world is strong and morally defensible. A cynic might question why it has taken bioengineered rice to arouse public awareness about Third World vitamin deficiency. Is genetically modified rice, patented by a transnational seed company, the best way to reduce vitamin A deficiency? Could consumption of a vitamin A-enhanced food source (other than beta carotene) put some people at risk, for example, pregnant women since excessive intake of vitamin A is associated with teratogenicity in humans?[25] Are there natural sources of vitamin A that could be introduced into the agricultural system?

There are ethical concerns regarding the use of Third World peoples as a testing ground for GM products. Political economist Robert Paarlberg has argued that developing countries have much more to gain from the GM crop revolution than do developed countries, and that because of their circumstances they should be willing to bear more of the risk for GM crops than the United States and Europe, where the regulatory thresholds are understandably higher.[26] This polarization among good-intentioned people who assess the risks and benefits of GM crops differently could be sensibly resolved by having an independent international agency such as the World Health Organization or the Food and Agriculture Organization consider the potential benefits and risks of a strain of rice that has been genetically modified with beta carotene.[27] The market system, operating through large and impersonal seed distributors and rice importers, would neither ensure democratic participation in a nation's choice to adopt GM seeds nor see that sufficient attention is paid to the human health, socioeconomic, and ecological effects of the adoption.

An ethical approach that gives primacy to autonomy must adopt as a starting premise that the populations who agree to be the early consumers of GM products are fully informed of the options and give consent to be part of the experiment. The consequentialist approach to GM crops/foods is based on the presupposition that the products are not inherently good or bad but should be assessed on health criteria and the unique sociocultural values of a nation.

Issue 8: Are there religious and/or dietary-ethical concerns about GM foods?
We live in a society of many cultural and religious beliefs concerning food. Some Jewish groups do not mix certain food types in the same meal, such as milk and meat products. These observant groups have

taboos against other food types such as pork or shellfish. Hindus and some Adventists do not eat meat. Vegans do not eat meat, fish, or eggs. How can we protect such beliefs within the context of GM foods? Would it matter to individuals who observe dietary rules that the gene from a food product that is a taboo in their culture is transplanted to a food product that is generally accepted? Would a vegetarian be opposed to corn that has been modified by the addition of a gene from an insect? Would an observant Hindu object to eating a plant into which a gene from a cow has been transferred? Would religions that oppose cannibalism object to eating animals with human genes.

Some companies have argued that a recombinant gene from an animal inserted into a plant is not the same as eating the animal. Typically, the animal-derived gene in the plant is expressed; otherwise, what is its function? That means that the person consuming the plant is consuming the protein that is found in the animal. If the person has a taboo against eating the animal protein, would that extend to the plant, which has been transformed with the gene (and its expressed protein) from the animal?

This is a question that must be answered by different religious and dietary-sensitive groups. The answers may not be the same. Suppose that the gene transferred from the animal to the plant is not expressed in the plant. Would that make a difference? Or, perhaps the gene in the animal and the plant codes for a similar (if not identical) enzyme. Does the fact that there is chemical homology between the foreign gene and its expressed protein within animal and plant affect the ethics of the discussion? Let us also suppose that we transfer a gene from an animal to a plant that codes for a nutrient, such as a vitamin or an amino acid. Would there be religious or ethical opposition to groups with special dietary considerations in these cases? John Fagan argues: "Although genes for proteins that are common to both plants and animals are related, there are significant differences in the information contained in those genes. That is, the cow hexakinase gene is different from the tomato hexakinase gene in information content."[28]

Is labeling a sufficient consolation for people who oppose the transfer of genes from a species they do not consume to one they do? Do people have a right to protect certain foods from being transformed by DNA from other species regardless of whether such products are labeled and regardless of whether they are found safe to eat? The issue might be viewed differently by people who follow dietary laws if the food supply was diverse enough to contain both GM and non-GM products and if they were distinguishable in both primary foods and processed foods.

Food security has taken on a new meaning in the last decade of the twentieth century. The discovery of mad cow disease in England and its spread to France and Germany has severely shaken European societies' confidence in the food supply and caused food-importing countries to be on high alert for contaminated beef and beef products. During the same period, the American public has been warned of the risks to children of pesticide residues on produce and of *E. coli* bacterial contamination in hamburger meat.

Also in the 1990s, a small group of transnational companies helped to define U.S. federal policies on genetically modified crops that circumvented public attitudes.[29] The confluence of mad cow disease, chemical contamination of food, and GM crops proved to be a recipe for heightened public skepticism against any dramatic changes in conventional food production. Concerns about food safety rekindled a deeper debate over the ethical beliefs underlying the production and distribution of food. Among the more audible voices in this debate are those who consider food production part of a vast network of players and stakeholders, including seed manufacturers, growers, distributors, primary and secondary processors, chemical companies, and consumers. They see farms operating within a larger ecosystem and argue that both must be protected for future generations. With issues looming like global warming, agricultural waste contamination of water supplies, and the spread of antibiotic resistance, we can no longer afford to look at the farm as an isolated system.

A primary ethical responsibility for food contamination became a legislative mandate in the United States through the enactment of the first food and drug law nearly a hundred years ago. Currently, food ethics has expanded beyond food toxicity to consider the methods of production, the stewardship of land, the treatment of animals, and the nutritional quality of food developed under modern methods. And now our deepest assumptions about the nature of food are being challenged by the transformative techniques introduced in plant biotechnology. These debates are creating new fault lines within the public interest community, forcing food security groups and environmentalists to reexamine traditional ethical principles regarding food that will cause them to either embrace or oppose bioengineered crops, until the political landscape opens up new areas of compromise.

Chapter Two

Why Food Biotechnology Needs an Opt Out

Paul B. Thompson

Aunt Orva

I'd like to introduce you to my Aunt Orva.[1] Now nearing eighty, Aunt Orva grew up in a small town in rural Indiana during the Great Depression. It was a time when people were not always sure where their next meal would come from, and Orva's family took food quite seriously. They were a deeply religious family, and took each meal as a gift from God. To this day grace is said in earnest at Orva's table, and the prayer of thanks is never slighted or given perfunctory treatment. Every meal is a religious experience at Aunt Orva's house, and nothing is put on the table that is not worthy of her faith.

Aunt Orva has never taken to Hostess Twinkies™, Hamburger Helper™, frozen TV dinners, or any other prepackaged, precooked, or pre-anything product when it comes to food. She's never eaten at McDonald's or Pizza Hut. Orva eats green beans, tomatoes, and ear corn in season, and the rest of the year she mostly eats what she has been able to can the previous summer. She will occasionally buy whole fruits and vegetables from the fresh, canned, and frozen sections of the supermarket, but never a can of soup or a frozen pie. I've occasionally seen a box of corn flakes or a loaf of store-bought bread in her cupboard, but whenever I've eaten at Orva's table, she has always served fresh-cooked oatmeal and fresh-baked biscuits or dinner rolls.

Orva complains about the quality of flour and yeast she gets these days, and she would never think about using a prepackaged mix. She longs for the days when she could get fresh milk and eggs from nearby neighbors. The one "new" product that she has welcomed in the last decade is organic brown-shell eggs, not because they are organic, but because to her it was like going back to the quality she knew in years gone by.

Like a lot of rural Hoosiers, Aunt Orva is a rock-ribbed Republican. Her conservatism comes out most forcefully when anyone interferes in her day-to-day way of life. Debates about economics, foreign affairs, and international trade pretty much pass her by, but the fact that she can't get locally grown tomatoes at the IGA is a serious insult to her sensibilities. (Fortunately, she can still grow a few tomatoes herself, and she can get what she needs for canning from the farmers' market.)

Aunt Orva and many of her neighbors are pretty sure that all this interference in their day-to-day life is a sign of serious moral decay. She's not the type to stand up and say that godless communists and the Trilateral Commission are violating her rights, but I sense the resentment she feels. At no time is that resentment more apparent than when she feels that her religious faith is being threatened. That happens mostly in two places: when the county commission threatens to end the nativity scene at the local courthouse, and when her ability to praise God through cooking and eating is compromised by shoddiness and impurity.

Our Food System

The food system has been under continuous evolutionary change throughout human history, yet the era of biotechnology is different in certain key respects. Throughout the centuries when farmers selected the best specimens for saving seed, when travelers introduced into local cuisine new varieties brought from afar, when plant breeders adapted crops for desirable traits, and even when food processors began to utilize artificial additives for flavor and preservation, there was always a way for people like Aunt Orva to keep her faith. Although it would be pretty natural for such people to say that all of these changes can't be good for us, faith is not really about food safety or health. One could say it is a matter of personal and cultural identity. People could say that food and eating tie them spiritually to the material world, or that the act of taking nourishment is, for them, a daily sacrament, a religious duty. They know that "substandard" foods will sustain a person in a bodily sense, but a practice of resolute care in the selection and preparation of food (and a thankful attitude) can sustain the soul as well.

It shouldn't be too hard to guess how Aunt Orva feels about genetically engineered crops or about the use of genetically engineered microbes in food production and processing. Truth to tell, though, she wasn't particularly outraged when she first heard about it. She assumed that genetic engineering would only be used in the kind of prepackaged foods that she avoids anyway. It was only when she heard that biotechnology would affect foods at the level of the seed itself that she began to worry.

The news that the entire nation's seed corn was affected put her in a real stew. It's not that she's worried about her health; she ate lots of nasty stuff growing up in the depression years and doesn't feel the worse for it. For her, the whole foods from which she reverently and painstakingly constructs her daily diet have now been contaminated— not necessarily unsafe, but reduced in purity somehow and less fit to put on the table. For the rest of her life, the daily ritual of preparing, blessing, and consuming food will be a bitter reminder of the way that her world has been shattered by the heedless acts of godless bureaucrats.

Biotechnology as a Moral Affront

The development and application of rDNA techniques in food and agriculture have obliterated the preparation and consumption of food as a sacred practice for people like Aunt Orva. Far worse than the banning of prayer in schools or public religious symbols, this is an invasion of their most private spheres—their home, their dinner table, their very body and soul. The disregard and disrespect that has been shown to the spiritual dimensions of food is among the most serious ethical issues associated with the introduction of rDNA techniques for modification of agricultural plants and food production technology.

It is disrespect of values that some would call religious, others cultural, and still others aesthetic, but in every case they are values that are extremely important to a person's ability to maintain a sense of constancy, identity, and faith in their daily practice. Threats to these values heighten one's sense of vulnerability and destabilize the nexus of presumptive beliefs that give coherence to daily habits. In disturbing the implicit order of day-to-day life, acts that threaten such values are associated with a pervasive feeling of foreboding and risk. To upset a person's capacity to act on such fundamental values without very good reason is the quintessential form of disrespect. By not labeling, disclosing, or even discussing the new biotechnology-based order of food products, purveyors of these new commodities threaten a way of life.

The only excuse that can be made for the industry, academic, and government decision makers who have perpetrated this affront is that they have not actually realized what they were doing. I think that this is, in fact, the case. Though I have been writing and speaking about this moral problem for over a decade, the reception has been mixed. I'm not personally opposed to eating the products of biotechnology, and I think that rDNA techniques hold out great hopes in agriculture and food production. So perhaps my own writings on biotechnology mix the good and the bad, and my previous statements of the disrespect that has been shown to those who don't want biotech foods may not have been as pointed as they should have been. For their part, critics of biotechnology have done a very poor job of stating the moral issue, and I will examine why this has been the case.

My own viewpoint is uncategorical: The affront that is being perpetrated on people like Aunt Orva is inexcusable, and it should not be allowed to continue. Before giving the ethical reasoning that supports this judgment, I should come clean on a few things myself. I don't actually have an Aunt Orva. She is a construct of some people I know here in Indiana where I now work and teach, some people I knew in Texas (where I taught for seventeen years), and a few of my real relatives. My Nana (born in 1910 in Oregon County, Missouri, and still living in Springfield) comes close in certain respects. She has probably never eaten at McDonald's or Pizza Hut, and she continues to grow her own tomatoes, even at ninety-plus. But like a lot of once-poor southerners her age, she liked the New Deal and loved Harry Truman. I don't think she ever forgave the Republican Party for Reconstruction, even though she must have only heard about it from her own grandparents. More relevant for our purposes, she *does* buy a number of processed and semiprocessed foods from the grocery, and to this day I rummage her refrigerator on my annual visits and discard out-of-date bottles of salad dressing and other condiments. Her ethic of not wasting food supersedes any concern for food safety.

Her preferences for the old ways with food are born of tradition, culture, and aesthetics, not some New Age food fad or concern for health. Like the fictional Orva, my Nana is none too thrilled with genetic engineering in her food, and also like Orva, she's not likely to write any letters to Greenpeace or the U. S. Food and Drug Administration (FDA) to express her opinion.

As for myself, I was raised by a working mom who valued convenience at the daily supper table. My mother invested a great deal of care and reverence into a few symbolic meals a year, and it pained her when the aesthetically, culturally, and spiritually correct foods could not be put on the table at those special occasions. But while we always said grace at my house, my brothers and I grew up in a household where we all constantly tried new things. My favorite meal as a kid may have been frozen fish sticks, canned spinach, and macaroni and cheese from the famous blue box. As a result, I'm not particularly bothered by the use of genetic engineering to produce crops that I'll eventually eat. I'm one of only a few Americans who actually got to eat one of the Calgene anti-sense Flavr-Savr® tomatoes, and I thought they tasted great. I'm as willing to say grace over virus-resistant squash or Bt corn as over my Nana's biscuits, gravy, and home-canned green beans. When I say that biotechnology is a moral affront, it is not that I am personally offended, either aesthetically or culturally, and it's certainly not that I am worried about my health.

Food Choice and the First Amendment

As currently practiced, biotechnology is a moral affront because there *are* plenty of people out there like the fictional Aunt Orva. She may not represent a majority of food consumers, but America is supposed to be a land where deeply held religious and personal values of minorities are preserved and protected. In addition to Orva, there are a few Jews and Muslims who interpret their dietary law to forbid the use of genetic engineering (though most do not). There are also plenty of former hippies, hypochondriacs, and New Age foodies who combine cultural, political, and aesthetic beliefs about what counts as good food with some idiosyncratic views about what's healthy.

The First Amendment articulates the moral principle of religious tolerance and free speech. It protects the right not only to engage in religious practices, but to frame and articulate one's abiding philosophical beliefs and principles according to one's own conscience. Beliefs about what one should be allowed to eat are certainly among them. Given the way that dietary practices form components of the faith for many of the world's religions and the overwhelming number of testimonials from people who object to genetically modified organisms (GMOs) on quasi-religious grounds, the current posture of the U.S. government is shocking. Working under the Coordinated Framework for the regulation of biotechnology, the United States is nearing the point where dietary preferences must be supported by peer-reviewed science in order to have any kind of legal standing at all! This is "science-based policy" gone berserk.

How did we get in this mess? In the next two sections I discuss two things that have conspired to produce the current corruption of our public policies concerning biotechnology-derived food. One concerns the framing of these food policies in the language of consumer choice. This rhetorical device has the unfortunate consequence of distorting the underlying ethical rationale for protecting the autonomy of those who want to opt out of the new biotech diet. The other is the tendency of biotech critics to ramp up the food safety issue to the point that biotechnology's challenge to cultural integrity and individual consent is overshadowed. When run-of-the-mill greed, concern about legal precedents, and shortsightedness are thrown into the mix, the result is a policy stalemate that has only gotten worse over a decade of increasingly acrimonious debate.

Not Consumer Choice

Many of those who have written about the ethics problem described earlier have used the unfortunate language of "consumer choice."[2] As seen by "choice" advocates, the problem is that consumers deserve a choice. When biotech foods, GMOs, or whatever one wants to call them, appear unlabeled on the shelves consumers have been denied their "right to choose." Now certainly, the story about Aunt Orva might be interpreted that way, but the language of consumer choice is an unsuitable vehicle because it can mean any of several different things. The ambiguity that is introduced into the policy discussion with the language of consumer choice has allowed those who wish to skirt the issue to do so. So how is consumer choice misinterpreted?

First, there are two ways of taking food choices to be important. One is based on the view that food choices involve rights, while the other attaches no ethical significance to rights at all. We'll start with the non-rights view of choice, common among regulators and consumer economists. From this viewpoint, it may seem as if we are only interested in satisfying preferences, and this leads in turn to surveys of public opinion and finally to an unacceptable kind of majoritarianism. If we were to take a survey, we might find that most people like to go to church on Sunday, but this would not provide a justification for policies that prevent people from going to church on Saturday. Even if a majority of people were just dying to eat genetically engineered foods, that fact wouldn't justify policies or practices that make it difficult for people like Aunt Orva to eat something else.

There is also a subtler problem. The policy wonks who work on consumer choice often start with the assumption that foods are solely of instrumental value. They think that foods are tools to produce healthy bodies and pleasurable experiences. By extension, choices among foods are useful only to the extent that they aid consumers in reaching these further ends. From this perspective, offering an option with no nutritional or sensory benefits is not offering a choice at all.

If one accepts the proposition that biotech foods are equivalent to traditional, nonbiotech foods with respect to nutritional and sensory benefits (a point that some critics contest), then no valid purposes can be served in offering people alternatives to biotech foods, especially if there are costs associated with doing so. Thus, those who start from these premises reach the conclusion that promoting consumer choice is actually a scam. "The advocates of consumer choice are trying to pick consumers' pocketbooks by alleging an implied benefit that simply does not exist," say the policy wonks, "then charging extra for it." Since segregating foods onto biotech and nonbiotech shelves can raise costs on both sides, someone who looks at the issue in this way is understandably skeptical.

The problem with this view is that someone has to decide whether nonbiotech foods produce any nutritional or sensory benefits. One possibility is that the regulators and the food industry decide this. The other possibility is that consumers do it themselves. Now there *is* a legitimate philosophical issue here. For example, one might argue that advertisers should not be able to say that their product tastes better than that of a competitor when scientifically controlled studies show that no one can tell the difference. Certainly, one can argue that claims to provide nutritional benefit ought to be supported by scientific tests. One might move from these arguments to a more general claim that the decision about whether to label so-called GM foods should be "based on science." Many people who say this think that they are preventing the unscrupulous from alleging a health or sensory benefit where none can be demonstrated under controlled conditions.

Personally, I'd like to know about those experiments, and if nothing more serious were at stake, I'd be happy for government policy makers to rely on them when deciding what sorts of options I ought to have. The rebuttal to this line of argument is to point out the existence of people like Aunt Orva. Unlike me, Orva is not particularly impressed by controlled scientific experiments. The First Amendment protects her right to live her life—and that includes her diet—according to a host of beliefs that are *not* supported by controlled scientific experiments.

The point can be made by comparing my case with Orva's. I'd like to know about the science so I can make better choices, and it's annoying and inconvenient to me when dubious claims are made. In Orva's case, her constitutional right to practice deeply held religious and cultural beliefs is at stake. It's crazy to think that my annoyance should trump Orva's right to live according to her religious beliefs! In this comparison, I lose. So do the eggheads who think that they can tell Orva what's good for her, because there *are* rights at stake here. One problem with the language of choice is that it has been too easy to slough off the rights issue by talking about benefits and costs.

Yet rights talk can be easily abused. If "consumer choice" is alleged to imply that people have a right to several options when making any given food choice, the claim of a "right" cannot be supported whatever initial plausibility and appeal it might have. Certainly, all of us would like options when we make any kind of consumer purchase, but what we would like and what we can claim by right are often two different things. If I walk into my local burger joint, I might like to have some options to the single-, double- and triple-patty combos, which I can order with or without cheese. I might prefer a soy patty or a fish sandwich instead. But I would be mistaken if I thought that I had a right to such options. Having a right to soy or fish burgers would mean that there must be some grounds on which I could claim that the proprietor of Burger Town is legally, morally, or at least customarily expected to provide these options to people on demand. Rights imply duties.

In general, rights imply that one can make a valid claim on behalf of the rights-holder. "Claim" here is understood broadly to mean that some good must be delivered to the rights-holder, or that the rights-holder must not be prevented from engaging in some particular practice. Claims that are valid under the law constitute legal rights. Claims whose validity rests on morality, ethical principles, or ordinary custom may not be legally enforceable, yet they still are routinely expressed in terms of rights. For example, if I'm next in line at Burger Town, I do have a legitimate expectation of being served next. I could, therefore, say that I have a right to be served next, and would be justified in the umbrage I would feel if the counter person allowed someone to jump the line. I may not have any legal recourse, yet as long as we are in a country (such as the United States) where queuing for service is the custom, it is reasonable to claim that I have a right to be served in the customary prescribed order.

The legal and philosophical literature on rights is huge and complex, and one must suppose that every time one person claims to have a right, someone else has disputed it.[3] But, honestly, is there any sense in which a person who arrives at the Burger Town counter can claim more than disappointment when the preference for a soy or fish burger is frustrated? One must either live with the standard burger options, or walk away hungry and unhappy, but *not* legally, morally, or constitutionally wronged. On the other hand, a person's rights *would* be violated were he or she prevented from walking away and forced, either by coercion or deceit, to eat the standard burger. Being literally forced to eat a burger is hard to imagine in any sort of normal circumstances; but sadly, deceit is easy to imagine in the American food system. A vegetarian who requests a soy burger and is knowingly served meat instead has been the victim of profound disrespect, as is the observant Jew who is slipped a pork patty, or the ordinary consumer whose burger is laced with horsemeat or spat upon by a surly counter attendant. (At one major outlet, french fries were allegedly "improved" by secretly adding beef flavoring.) In any of these cases, people could say that their rights have been violated. But the guy who orders fish and is told, "Sorry, not available"? He may go away hungry, but he never had a right to fish in the first place.

This little scenario is relevant to biotech food. When the language of consumer choice is interpreted to mean that a shopper in the local supermarket has a right to something other than GMOs, we are arguably in the same position as was the deceived fast-food consumer. Most grocers happily supply a huge array of products, and Americans have come to expect this. But if a given grocer doesn't want to carry beer, or cigarettes, or *Playboy* magazine, can shoppers claim that they must be provided with the opportunity to buy these goods, and that their autonomy will be compromised if they are unavailable? If the merchant can choose not to carry *Playboy*, can't he or she also choose not to carry organic milk, free-range eggs, or margarine? There are still a few rural grocers who won't.[4] And what's morally or legally wrong with operating a store that carries no milk or eggs at all? Have shoppers at the milkless grocery been victims of a rights violation? So given all this, why should we think that consumers have a "right" to non-GM foods?

Of course, what would be problematic is the situation in which consumers are prevented from going elsewhere to find the products—whether milk, non-GM foods, or *Playboy*—that they want. They may have to bear some inconvenience or even pay a premium to get these products. But unless the inconvenience is so great that it can legitimately be called coercive, it shouldn't be said that anyone has violated their rights. If they are locked into a system that denies them every opportunity to act on their values—if the local Mafia enforcer is insisting that they shop at the milkless store, for example—then their rights are being violated. It is the circumstance of being forced into a system in which there is no exit, no way to opt out, that is objectionable, rather than the fact that some desired alternative is unavailable. The issue is not that particular merchants have a duty or obligation to provide a GM-free alternative, but that the whole food system forces people like Orva into circumstances they find intolerable.

Exiting the System

To put this point into legal and philosophical terms, the problem arises when people are denied exit, not when they lack some particular option. Having exit means that one can opt out of a given set of arrangements. The idea derives from social contract theory. Government is thought to rest upon an implicit agreement, on consent of the governed. People agree to be bound by laws and authorities with which they do not agree in every particular. The sense in which people "agree" to this is that they can "opt out" of the social contract, though in so doing, they forfeit the benefits conferred by government, including the protection afforded to person and property. In the practical sense, this means that civil societies must never foreclose the right of emigration. Citizens can be said to have consented to the social contract only if they have some possibility of exiting society and its laws. The fact that people stay can then be interpreted as implicit agreement to be bound by laws and civil authorities.

Exit is a minimal requirement for consent, and many would argue that exit is far from a sufficient condition. Work on informed consent in medical ethics demands much more rigorous requirements of information and presentation of options. The issue in that context is the imposition of known physical risks. The social contract discussion is especially relevant in the present context because freedom to practice one's religious convictions has been thought to be one of the main reasons that people would quit one society and choose another. America was established as the land where those exiting European societies with established churches would be free to practice their own faith. What is critical in this context is that without exit, a citizen is forced to practice the state religion. This is what constitutes the moral affront. The right of exit is a necessary condition for a just society. Without it, civil societies are coercive and repressive.

In emphasizing exit, it also becomes clear why novel food technologies differ from traditional crops and methods. If Orva claims the right to non-GM food, why can't I counteract her claim by claiming a right to the tasty, and cheaper GM foods for myself? Why should one person's right to have X hold precedence over another person's right to have Z? Indeed, why?

Can I, however, plausibly claim a right to exit from the situation of eating conventional or traditional foods? I've been eating them all my life and it never bothered me before. I have not established a record of protest and resentment against this practice. There is absolutely no evidence that I believe myself to have been a victim of coercion and discrimination because I have spent my life eating non-GM foods. Yet the sense in which Orva feels herself to be a victim of coercion and deceit is pretty obvious. Even those of us who don't give a flip about the GM controversy can empathize with Orva, and we can understand why she wants to opt out.

To conclude this point, then, the "consumer choice" argument makes a weak claim when interpreted in the language of benefit-cost trade-offs, rather than rights, and a dubious claim if interpreted to mean that someone can rightfully demand that society provide them with a particular product. However, when the rhetoric of consumer choice is interpreted as a matter of exit, we can see how it involves an important moral claim.

One implication is that people like Orva must have some way to avoid GM foods. This is not, in itself, to say that GM foods must be labeled. Arguably, the U.S. rules of organic certification could be said to satisfy Orva's right of exit. To be labeled "organic," a food may not contain GM ingredients. There is still much to say about labels, organic standards, and other policy instruments that might be proposed, and I cannot cover all these issues in this chapter.

My point has been to establish the ethical terms in which these policy instruments must be debated. They should not be debated in terms of their costs and benefits. One should not get into a debate where Orva's right to eat non-GM is seen to be in conflict or tension with my right to eat GM (or the right of some starving Third World person to eat GM). Labels, standards, and other policy instruments should be debated in terms of their ability to secure exit for people like Orva. That is the ethical bottom line, and it is a minimal condition for an ethically just food system. If it truly is impossible to secure exit for Orva in a food system with GM crops and processes (which I strongly doubt), then so much the worse for GM!

Not Food Safety

In constructing Orva, I've made a point of noting that her concerns about GM foods do not derive from a fear that these foods will make her sick or cause harm to anyone in a medically certifiable way. But the critics of biotechnology tend to promote the idea that you should be worried about GM foods on grounds of safety, and they put forth three kinds of arguments to support that contention. One stresses the possibility of allergic reactions that might occur when known allergens are inadvertently introduced into a "safe" food through genetic engineering, or when totally novel proteins are introduced into diets. The second stresses uncertainty: How can we really *know* that these foods are safe to eat? The third type of argument is negative and deals more with nutritional quality than with safety in the narrow sense; it rebuts claims made by the boosters of biotechnology, who promote products such as Golden rice, which will allegedly improve the nutritional quality of rice by adding a precursor of vitamin A.

These arguments all present entirely legitimate points to debate, but my claim is that they are much less relevant to issues of choice, consumer rights, exit, or labeling than is commonly supposed. If critics were successful in convincing the FDA that a given product was unsafe, either because of allergenicity or because of uncertainty, the product would not be allowed on the market. It's that simple. FDA scientists themselves can, in my view, be relied upon to make such judgments.

FDA scientists don't need to be convinced by outside critics, because they themselves will bring the critical arguments to bear.

On the face of it, the ethics of food safety are fairly straightforward: If there is any legitimate doubt about the safety of a product, it is unethical to allow people to eat it. I think the FDA tends to see the issue this way. The FDA has no doubts about the safety of products currently on the market, and that is why it has taken the posture it has with respect to GM foods. If there are doubts, labeling is an inadequate response. Keep it off the market entirely until those doubts are resolved.

But of course, there is more. There are plenty of people who are like my Aunt Orva in that they don't trust government. They take a jaundiced view of what present-day science tells them about what to eat. Unlike Orva, they are less concerned about religious and cultural meanings of food, and very concerned about safety. They've seen too many revisions of dietary and food safety advice based on "the latest scientific results." They've lived through or read about four decades of debate over pesticides, and seen products once promoted as "good for you" withdrawn from use because of their toxic properties. So, like Orva, they may be claiming a right of exit from GM foods, but unlike Orva, their concern really does relate to safety.

Now, the First Amendment probably protects a person's right to believe whatever he or she wants about food, health, and the effect of diet on the body, just as it protects the rights of people to handle snakes and forgo medical treatment for religious reasons. But in this context, we are talking about people who do not have religious or even cultural reasons for their concerns, and whose motivating concern truly is the safety of their food.

Furthermore, we have at least a century of government stepping in to regulate claims that are made about the efficacy of various potions, procedures, and dietary regimens. The point of this regulation is to protect the public from charlatans and snake-oil salesmen who exploit our insecurities and our general ignorance about medicine and health. Though the people at the FDA are rather circumspect in their public statements, they may well regard the critics of biotechnology as charlatans peddling an agenda of self-promotion or some political end unrelated to safe food. "If the critics can worry enough rubes about the safety of GM foods," the regulators may think privately, "they can build their base of contributors and build support for anti-agribusiness policies that will weaken free trade or help small farms. But we at FDA are not concerned with such political issues. We are concerned with actual risk."

There is evidence to support this jaundiced view of at least some of biotechnology's critics, but many critics are quite sincere in expressing concerns about the safety of GM foods. Their sincerity suggests a few questions: Do people have the right to substitute their own beliefs about safe food for those of the FDA scientists? Do critics have a free speech right to persuade people that GM foods are risky? In fact, the moral issues in the domain of food safety are far from black and white. The shades of gray in play are numerous and difficult to sort out. Even if some critics are sincere, is it ethical for someone who is not particularly concerned about the safety of GM to ramp up safety concerns as a way to gain allies for an assault on agribusiness? In any case, should the FDA make policy decisions based solely on the fact that some people take themselves to be at risk, even in the absence of any evidence of risk?

I acknowledge that these are important questions, but discussion of them would broach issues that take us far from the debate over labels, consumer choice, and the right of exit. Yet one can find people who have food allergies, or who have a close friend or relative with food allergies, who do assert a right of exit from GM foods. Let's call our candidate Edna. Edna may be genuinely concerned about the safety of GM foods, but her argument for a right of exit would look somewhat like the one I have sketched for Orva: "I don't care what the scientists say, I have the right to avoid any food that I think I might be allergic to." The strength of this argument, however, is not that an alleged risk to health lies behind it, but that people should never be put in a position where they are effectively coerced into food choices simply because they lack scientific evidence to support their preferences. Edna doesn't need to have a particularly good reason for avoiding GM, any more than she needs a good reason for avoiding strawberries, chocolate, or soybeans. If she thinks that any of these foods might make her ill, she ought not be put in a position where she is forced to eat them. Ironically, as long as we see Edna's argument as parallel to Orva's, it's a strong argument. If the fact that Edna's argument involves a health claim is overstressed, it becomes weaker. Let's see why.

As a health claim, Edna's argument is weaker than Orva's because then it really does seem to go up against science, where it is bound to lose. Edna can believe what she likes, but science provides a more robust and better-founded set of considerations for making judgments about safety and health than do idiosyncratic beliefs. What science says at any given point could be wrong, of course, and Edna could be right. Science is fallible. But if Edna's reasons for thinking what she does have any currency or validity, they should have been incorporated into the existing scientific consensus.

Of course, more shades of gray are lurking below the surface here. Are there forces at work, such as the corrupting influence of corporate ties, that might distort the scientific consensus? If Edna thinks that there are such forces at work, she may discount what scientists say. Edna has every right to do this, in my view, but such discounting is a move away from a direct health argument. It is a move to more culturally based considerations, to considerations that resemble Orva's more than they resemble those of a scientific argument about diet and health. In this case, they are considerations grounded in Edna's views on human nature, and on the way that science is institutionalized in capitalist societies. Like Orva's religion or her Republican suspicion of government, Edna's doubts about the integrity of scientists do not refer to a property of GM food that would make it unsafe to eat.

A philosopher would say that these are second-order beliefs that bear on Edna's willingness to credit the scientists' claims about food safety, but in and of themselves, these beliefs don't tell Edna (or anyone else) anything at all about safety. The First Amendment protects the political beliefs someone would form about such issues as surely as it protects religion. So again, Edna can claim a right to opt out, but the tie to health and safety has now become rather indirect. It's not safety, but culture and liberty of conscience that forms the basis for her exit.

Conclusion: What to Do?

The cool and analytic treatment that I have given the issue here probably won't persuade anyone who is on the brink of taking up arms against genetically engineered food. Don't expect activists to start using this kind of First Amendment/liberty of conscience argument to support their critique of GMOs. For their purposes, it's much more useful to have a raging public freaked out about the safety of GM food, or at least saying that its choice has been denied. Unfortunately, this creates a situation in which regulators and industry representatives can rest comfortably in the opinion that they have a more subtle and nuanced understanding of the issues. "Only we truly understand the costs that would be borne in any attempt to create an option to GM foods," they say to each other over coffee. "Only we understand the subtleties of risk arguments and the way that scientific evidence bears on the question of allergies, nutrition, and food safety." The pro-biotech crowd can even portray themselves as courageous defenders of rationality, defending reason from the onslaught of radical organizers, postmodern liberal academics, and an irrational mob.

That's balderdash. Far from being defenders of liberty, they are perpetuating a moral affront and contributing to an erosion of one of the most fundamental liberties on which the American republic was founded. People must not be required to produce a risk assessment that supports a deeply grounded set of philosophical beliefs about what they should eat, or about what they should feed their children. Now, the motive for appealing to science may be well intentioned. I would agree that people should pay attention to science, but when I say that, I'm giving prudential advice. Interested in healthful food? I say, "Well, here's what the science says." I'm certainly not stating a principle that government and industry can use to rationalize a transformation of the food system that vitiates liberty of conscience when it comes to diet. The opt out is a necessary condition for protecting the right of exit and, in turn, the right of citizens to lead lives based on religious, cultural, and even political convictions of their own choosing.

It matters a great deal that we are talking about food. Anthropology will reinforce the importance of food as a carrier of culture, and the complex significance that people of both primitive and modern societies attach to matters of diet. The arguments I am making here would not apply to fundamental technical changes in the steel or auto industry, or to the way that utilities decide to generate electrical power. The food that people eat becomes part of their bodies. It is not surprising that people associate spiritual and sacramental meaning with food. As a result, the risk and environmental issues that would be associated with the evaluation of technology in the manufacturing or energy sector are augmented by fairly straightforward—if complex and even subtle—religious and cultural considerations when it comes to food. Liberty of conscience—religious freedom—is at stake in this debate, and anyone interested in the right of citizens to practice their faith should be worried about what's happening to people like Orva.

So what is important here is the protection of exit, and the provision of an opt out—a way to avoid eating foods that are contrary to one's fundamental values. As mentioned briefly already, it is possible that organic provisions already in place in the United States meet this requirement, though there is literature available arguing that they do not.[5]

I do not regard it as my place to propose or evaluate policy instruments that would protect exit. Someone more like Orva is in a better position to do that. We should expect that any proposal specifically oriented to exit would need a substantial amount of discussion and debate. My goal here has been to articulate why food biotechnology needs an opt out, and to articulate the strongest ethical and philosophical arguments in the clearest language I can provide.

The alternative, of course, is that the representations of "freedom of choice" arguments are willful—that industry, government, and academic scientists are fully cognizant of the ethical argument that is being made, and that they *choose* to misinterpret it so they can argue the point on economic or safety grounds where the case for biotechnology can be more easily made. That is the suggestion that gets us from Orva's concern to Edna's. Perhaps we ought to be resisting food and agricultural biotechnology because the people who have been placed in charge of it are just a bunch of untrustworthy weasels. However, I know a lot of these guys personally, and I don't believe they are weasels. I suppose that's the main thing that keeps me from gravitating toward Edna's viewpoint so far. I can understand perfectly why someone like Edna might go there. I respect her point of view, as I do Orva's, and I will fight for their right to live and eat accordingly.

Chapter Three

A Naturalist Looks at Agricultural Biotechnology

Brewster Kneen

Right and wrong, good and bad. In a secular, multicultural, and highly individualistic society, how can one begin to think about morality and ethics? Even if we abandon the terms right and wrong, good and bad, there remain the questions of doing or not doing, acting or not acting, and being responsible for what one does or does not do. But responsible to whom? And for what? And how and when should that responsibility be expressed? When a technology—such as biotechnology—just "arrives" or "presents itself" (as a journalist recently said in my hearing), without agency, then no one need take responsibility for it.

The result is that questions about the use or application of technologies, particularly in the area of human medicine, are relegated to the downstream issue of whether or not the object of the technological application has knowingly consented to be such an object.

In Western industrial agriculture, technological "advances" have seldom been subjected to any critique as to their effects or potential consequences, and farmers have been trained to view all new agricultural technologies as instruments of progress, to be accepted as such regardless of any personal misgivings. A breakthrough in this regard is now occurring as the public turns against genetically engineered foods and demands labeling. Labeling could be viewed as the mechanism (or "enabling technology") required to adhere to the practice of informed consent—in order to be able to avoid the products of agricultural biotechnology, genetically engineered foods.

Ethics and morals, however, are about more than consent to proceed with a technology already underway and commercialized. Should there not be questions about the process itself, its purposes and consequences, and the very attitudes it embodies? Should there not be questions about the consequences of these attitudes for the ways in which organisms and ecosystems as a whole are regarded and treated by the technology itself? Should there not be moral or ethical constraints to the reduction of life to a matter of mere technology, as in biotechnology, which reduces life to a code to be broken? This form of inquiry is neither new nor daunting. As Brian Johnson has observed, "Trawling society for moral standards and ethical views leading to informed debate is not a new process; we do it all the time in, for example, areas of medical development and animal welfare."[1]

I was brought up to respect others, that is, to let them be who they are, to give them "unmolested" space within which to be. These are the best words I can find to express an attitude that has been a constant in my life since childhood. Telling someone to "respect" others is not, of course, the same thing as telling a person to "like" others or to agree with them. It is simply saying, "Respect their *otherness*." This does not apply only to other human beings. It includes all creatures, plants, and even natural forces and elements, such as rocks and earth, wind and water, and sunshine.

Perhaps "appreciate" is a more appropriate word than respect, or maybe the two should be conjoined as respect-and-appreciate. The appreciation is for the complexity and wonder of life—more material, perhaps, than the "mystery" of life, which seems to make of life something remote and even alien.

One might well respond that our society is founded on the idea of respect for the individual, but respect and appreciation for others—for all Creation/creation—goes well beyond the Golden Rule. (I use "Creation" simply to indicate greater inclusivity than is generally meant by "nature.") Affirmation of an ethic of respect would, to my mind, rule out domination and control as acceptable, unequivocal pursuits, whether the control was exerted over wind and water, the earth and all the creatures in and on it, or other human beings. It would exclude killing other people, whether for personal gain or in the name of the state; it excludes seeing others as the means of my own personal gain; and it excludes considering everything and anything outside of myself as nothing more than a *resource* for my exploitation. Under this rubric, food crops and farm animals then are more than "raw material" or "commodities."

When I try to discern the common ethical thread of my life, what it comes down to is respect for all forms of life and for Creation itself. It is this ethic that has subtly informed my growing understanding and critique of biotechnology. I say "subtly" because it has never been a conscious measuring stick of what I think and do, and I can only now identify it in retrospect.

In wrestling with the practice of biotechnology, the assumptions on which it is based, and its implications and intents for life, I have become increasingly aware of the extent to which it is the project of a very particular and peculiar culture. It is a particular way of looking at the world that is unlikely to be shared by the vast majority of its inhabitants, now or in the past. Recognizing this has led me to formulate the ethic of respect and appreciation of otherness in terms of boundaries and interventions.

The most obvious boundary to be respected is the skin of another person or organism—or the bark of a tree, the bank of a river, or the shell of a nut or a turtle. The real boundary is less specific, however, for the organism requires space within which to move and grow. It requires space within which to modify its own internal otherness. It also requires space within which to interact with other organisms and its less well defined environment. The health of the organism requires respect for the boundaries of all of these spaces.

A healthy river, for example, requires more than the banks that define it in the dry season. It requires headwaters, floodplains, and estuaries. Rivers are violated when they are confined behind levees and dams, and are known to wreak their vengeance from time to time in retaliation. Unfortunately, the human victims of the river's vengeance are often the same marginalized people who were the victims of the boundaries imposed on the river in the first place. The real doers of the deed remain isolated and safe in their faraway offices and boardrooms.

Communities, societies, and ecologies also have boundaries that deserve respect and appreciation, and it is as important to respect and appreciate their complex configurations and compositions as those of simpler organisms. Plants, animals, and microorganisms have their own times and seasons for breeding and reproducing—their time/seasonal boundaries that are intimately related to their larger environment and not simply products of their own whims.

Boundaries are often invisible, or only subtly visible. William Cronon describes, for example, how the British colonists were unable to recognize the settlements of the New England native peoples because they were not fenced, and Hugh Brody describes the settlements and boundaries of the hunter-gathers of Canada's far north, boundaries that do not appear on any map or satellite photo and are known only to the inhabitants of those vast reaches of seemingly barren land. [2,3] The boundaries are nonetheless real and define the space required for the organism—in this case, the native peoples of the north—to thrive.

An ethic of respect for boundaries requires that false boundaries not be created—such as a boundary between a mother and the fetus she carries, or the boundaries imposed on geographies to suit colonial interests—in violation of the evolved relationships of ecosystems, including those of humans. The wonder of life is to be found in the fact that one life contains another, and that there is constant evolution of the relationship between one organism and another, from utter dependence to complete—at least bodily—separation. The wonder of life can also be appreciated in the intricate and delicate interdependencies between organisms and their environments, predators and their prey. [4]

Just as the recognition of boundaries can be an expression of respect and appreciation, interventions can be acts of overstepping the bounds for the sake of healing and restoration. Species can become invasive when they are carried beyond their natural boundaries into foreign lands and no longer contained by their traditional boundaries. As a result, it may become necessary to impose boundaries on the invasive species out of respect for the boundaries of the invaded habitat. The release of transgenic organisms into the environment can be regarded in much the same way.

Discernment of the intent and consequences of interventions on the one hand, and respect for boundaries on the other, is the stuff of ethics.

Morals and Ethics

Perhaps it is the inheritance of an earlier age, which assumed one could speak about a "Christian nation" or that the citizenry shared an ethic based on a common religious faith and European culture, that has left us at such a disadvantage in dealing with questions of ethics and morals in a multicultural, multifaith society. Now the problem is being compounded by the global campaign to shape this diversity into a market monoculture by the dominant powers of capital. Biotechnology plays its part in this campaign as a means of extending control over organisms and life processes for purposes of profit.

The apparent contempt for the integrity of organisms and ecologies that seems to be essential to the practice of biotechnology raises the question of ethics in a far more significant way than the downstream question of informed consent. To identify the ethical or moral principles that are missing, it is helpful to distinguish between ethics and morals. The following distinctions are not intended to constitute an academic or a theological exercise, but are a personal attempt to get a handle on the issues.

Morals are essentially the social norms or mores of a particular culture or society at any given time, their authority residing in the social consensus about how the world is to be regarded and what is to be our proper place within it, whether individuals explicitly consent to it or not. Only when a society recognizes and accepts a social code wherein the welfare of the society as a whole is prerequisite to the welfare of any individual within it, can one then speak of public good.

Given that morality, and with it the social or public good, is a social construct that provides meaning and value, it is only logical that the loss of social cohesion and identity leads to the loss of the idea of social good. The process might also be considered in the reverse: that the reduction of moral considerations to the strictly individual level leads inevitably to the erasure of the notion of public or social good and its replacement by the "good" of personal gain or benefit. Morals simply disappear, replaced by "whatever works," "whatever one can get away with," or "what the market demands." Social decay and injustice are among the obvious consequences. Genetic engineering is both a tool for and a product of this erosion of public good.

Margaret Talbot has pinpointed at least one of the ways ethics becomes subverted to a warped notion of a public good.

> Already, bioethicists who favor cloning have begun outlining the categories of people who might consider it. Indeed, for the last several years, those in the profession who have taken up the subject of human cloning seem to have been more concerned with identifying its worthwhile applications than with raising serious alarms about it. . . . Indeed, the pro-cloning bioethicists I talked to often resorted to a sort of

consumer logic: there's a market out there that wants this, and who am I to say they can't have it?[5]

What I am describing as the foundation or necessary context of morals is what Ivan Illich has described as the basis of ethics. The basis of ethics, he says, has always been *ethnos,* "the historically given 'we' which precedes any pronunciation of the word 'I'." Ethics, says Illich, were formed within an ethnic boundary that gave them shape and substance. They applied only within the boundaries that defined a given people in a given place with a given tradition.[6] I do not contest Illich's definition, but in a multicultural society I think it better describes what I choose to refer to as morals.

Ethics then becomes those principles that gain their authority not from the *ethnos,* but from outside society, through reference to a cosmic force, a deity, or sacred spirits, and thus apply directly to personal behavior regardless of social norms or morality. To speak of ethics is to imply an absolute, a line of behavior one will not cross. Ethics is a matter of principle, not relativity or social tolerance or acceptability.

Where the line is to be drawn, not *whether* the line should be drawn, is the subject of ethics. Example 1: A scientist in the employ of a government regulatory agency is pressured to approve a drug that she knows does not fully meet the requirements. She refuses, on principle, as a matter of personal integrity, or perhaps because she does not want to be responsible for the consequent harm the drug could cause if put on the market. Example 2: A person believes in the sanctity of life that prohibits killing (conscientious objection), even though such a position is regarded by a state at war as immoral and illegal, with the result that the offender may be jailed or even executed. In the first example, the personal ethic runs counter to institutional mores of expediency, and the individual in that context has to decide whether to take a stand based on principle or to accede to institutional or employer demands. In the second example, the conflict is starkly between the power of the state and the conscience of the individual. Unfortunately, few seem willing to take this stand.

A third example is that of a tenured professor who is also a consultant to transnational corporations because of the particular, and valuable, skills he has in biotechnology. One day, however, he discovers the uses to which his work is being put and, realizing he has absolutely no control over that, decides to quit his field altogether, in spite of his love for the science. He simply cannot be party to what he realizes will be the outcome of the use of his "technology" by a corporation in search of profit. This professor has made an ethical decision.

If we move beyond anthropocentrism and include other animals and organisms in our circle of respect and appreciation, we could face the interesting challenge of evaluating our human activities in the light of the possible ethics and mores—or "sensibilities"—of their cultures, ones that we hardly even recognize, much less respect. As Richard Lewontin has said, "If one wants to know what the environment of an organism is, one must ask the organism."[7]

While moral and ethical standards may derive their authority from different sources, both include personal and social responsibility. In the case of morals, the individual is responsible to the society for upholding social standards and public good, while behavior in private is apt to be considered a personal matter not subject to the same rules. (Spousal abuse is an all-too-common example.) In the case of ethics, the person is responsible to the authority behind the ethical standards, for both private and public behavior. In both cases, the individual is held responsible for his or her public actions or inactions. Personal responsibility for *both* personal and social behavior lies at the heart of ethics.

(In some religious traditions, including Amish and Mennonite, disapproval of personal behavior, considered as antisocial, is expressed through "shunning," the exclusion of the offending individual for a period from the social life of the community. Similar practices can be found in aboriginal communities, which are now using this as a healing technique for addictions and family violence.)

The Trivialization of Ethics

The practice of biotechnology—the willful manipulation of life-forms and the deliberate and systematic violation of organisms and ecologies—raises myriad ethical issues, yet so far virtually all discussion about the ethics of biotechnology has been about the application of the products or the "technology" of genetic engineering. This is reflected in the regulatory philosophies of the biotech powers, in that it is only the product that needs to be regulated, not the process by which it is produced—if there is any regulation at all. Even the term "regulatory" has been replaced with "approval," so the process is now referred to as the approval process, in parallel with the exclusive presentation of the "benefits" of the "technology" to the exclusion of any negative real or potential consequences or costs.

The culture of biotechnology would appear to be utterly utilitarian, with no thought given to any ethical responsibility to the organisms affected or offended, be they plant, animal, or microorganism. As one of Canada's most notable figures in biotechnology policy put it, "Genomes are not people and technologies per se are not good or bad."[8]

The process/product distinction is essential to the construction of biotechnology and rests on the assumption that all technology is value-free, its moral or ethical character—if one can ascribe such to a technology—being derived solely from the use to which it is put. A handgun, for example, has a market value that bears no direct relationship to its possible use, even though its potential use, or the context of the market, may influence its market price. If the gun fires accurately when it should, then it is just another product on the market. Yet the gun itself has its own agenda: It was designed to kill. This obviously raises questions about possession of the gun in the first place.

Similarly, Bt corn is designed to kill the corn borer, but it is sold to the farmer, and the public, as being "insect resistant." The language is a deliberate deception: from the corn's perspective, it may be resistant; but from the perspective of the insect, and its predators, it is a poison. Not only is the character of the corn, and the technology (gun) in its hands misrepresented, but its social consequences—the effects it has or may have on its ecology, human and otherwise—are also ignored. Questions that ought to be triggered by the process of genetic engineering itself, since there is no way to know the broad, long-term consequences of genetic engineering, are not asked. This moral and ethical deficiency arises from the attempt to avoid agency and responsibility for its consequences.

The Ethical Canary is the title of Margaret Somerville's book on "science, society, and the human spirit." While the book is about her field of medical- and bio-ethics, the title itself is perhaps the most important comment on her treatment of ethics. Nowhere does she offer a critique—ethical or otherwise—of the "technologies," including reproductive and genetic technologies, xenotransplants, and the like. Her utilitarian and anthropocentric focus is entirely downstream on the ethical questions raised by the use of these technologies. For Somerville, the ethical questions that various practices raise are the canaries that "test the air in our societal mineshaft," but she does not ask if the canary gave its informed consent to being placed in the mineshaft; she does not ask who developed the mine or whether it should have been developed at all; she does not ask if there were alternatives to mining; she does not raise the question of where the tailings from the mining are going, and so on.[9]

Cultural Assumptions

Many people appear disinclined to consider the possible harmful environmental effects of genetic engineering, or indeed *all* the effects on the targeted organisms, be they plant or animal (herbicide tolerance, insect "resistance"), or on the ecology of organism and environment. Such a view reflects an attitude toward nature itself. To even suggest there may be ethical issues to be considered concerning the so-called technology itself may be a cultural leap for those immersed in the culture that has produced genetic engineering.

The rationalization of agricultural biotechnology implicitly rests on two cultural assumptions. First, that Nature itself is stingy, alien, and hostile, an enemy territory to be conquered and colonized. By itself, this assumption goes a long way in explaining why ethics are not applied to the construction and use of biotechnology. As in any war, killing, maiming, and mayhem are legitimized as essential to achieving victory over evil.

Second, that "natural resources," including "genetic resources," are both scarce commodities and inherently *deficient*. (A university biology professor startled me by announcing to our audience that "rice is deficient in vitamin A," which is like saying strawberries are deficient in protein.)

The scarcity ideology provides the foundation for the biotech industry propaganda claims that we will only be able to feed the world and save the environment through biotechnology. This logic deliberately excludes any consideration of the causes of malnutrition and starvation and of distributive justice as an alternative course of action to address the problem of apparent scarcity. However, if scarcity is a "man-made" problem, both in the sense of being constructed through the process of commodification for purposes of "the market" and in the sense of being the result of vast overconsumption (fossil fuels, water, topsoil, etc.) by a powerful elite, then the problem is social and political in nature, not technological. Scarcity, in other words, is socially constructed and consequently both an ethical and a moral problem; ethical because it is the consequence of those with the power to command resources assuming that it is their right to do so, and moral because it destroys the social fabric by deliberately creating competition for what is defined as scarce resources, including food. Agricultural biotechnology is then offered as the cure for the problem, albeit without addressing the causes of the problem.

The construction of scarcity amounts to creating a boundary around sufficiency and the commons in the name of protecting private property—much like building a wall topped by broken glass around the gated communities of the wealthy. There is a long history of this practice, of course, but it is being taken to new extremes, and to a new level of ethical violation, with the patenting of life-forms—genes and fragments of genes and genetic engineering processes that amount to nothing less than the privatization of life. Life itself—or perhaps more accurately, the building blocks and expressions of life—is being transformed into a commodity for sale to the highest bidder.

The implication that nature is itself deficient is expressed by the language of "improvement." Industrial agriculture has long spoken of "improved seeds" in referring to hybrids, but now the term is used to describe the products of genetic engineering. There is a constant stress on improvement, as if this comes only by the hand of humankind, who must make up for Nature's deficiencies. This becomes, of course, a license for whatever the technicians want to do.

The cultural specificity of this is apparent in the current conflicts over patent claims on traditional crops and herbal remedies between traditional societies that have known about and utilized these crops and remedies for ages, and Western business interests that seek to claim ownership over them and privatize them for profit.

The assumption that patenting and intellectual property rights are universally valid and acceptable is a reflection of the arrogance of the culture that holds this position. For most people throughout most of history, the commons has been a reality and a foundation of their social and economic life. To destroy or appropriate the commons for private gain amounts to the erection of a boundary very similar to those imposed by colonialism and legitimized only through force of arms.

Florianne Koechlin interviewed Ethiopian plant scientist Egziabher Tewolde on the subject of his position on patents on life. Tewolde is spokesperson for many African countries in international negotiation on patenting and trade.

FK: You're partly responsible that all African states refuse to accept patents on life in the ongoing TRIPS negotiations. Is this position not rather naive?

ET: Depends on which world you're talking about. Our world has no patents. Patenting living things is completely alien to our world. If we're going to accept this it is going to disrupt our systems of living and production systems based on biological resources, it will disrupt the whole of rural life. We have no choice but to fight and protect our interests. Not patenting life is central to these rights. If industrial countries want to keep patents among themselves, let them keep it, but we will not recognise patents on living things.[10]

Progress and Agency

Can we even dream, with any sense of justice or righteousness, of feeding the hungry with the bitter fruits of a profoundly violent system of forced labor from our crops and animals? When concerns regarding biotechnology are raised, the biotech industry is quick to label the questioner a Luddite attempting to block Progress. Progress is the way we overcome Nature's hostility and stinginess. In Western culture, it is a moral imperative, a socially mandated "good."

The idea of Progress itself rests on the foundation of technohistorical determinism, and it is this premise that excludes Progress from ethical consideration. In fact, technology is frequently treated as if it were immaculately conceived, "arriving" without human agency or responsibility. The scientists and technicians bringing it to fruition see themselves as functioning merely as midwives in a process in which they have no real agency and over which they have no control and hence no responsibility.

There is, of course, an overwhelming hubris to all this. The assumption that only our science can bring Progress, that our Progress is both desirable and inevitable, that we are sufficiently omniscient to actually understand what we are doing and to correct or compensate for any problems we might create (the technological fix) is sheer, overwhelming arrogance. In the eyes of many, it is precisely this arrogance that identifies the practice of biotechnology as unethical and immoral.

Just as there can be no ethics without a responsible agent in an acknowledged culture, there can be no technology without a context, including agency. There is an element of agency embedded in all technology because there is a designer with an agenda behind all technology. Genetic engineering carries this agency a step further in that organisms themselves are self-replicating active agents in relation to their future generations and their environment, in some ways not unlike nuclear "waste," which carries on with its own agenda long after it has been "disposed of."

No technology has ever arrived without an agent or existed without a context, without an environment or culture, from the simplest flint axes to today's nanotechnology, genetically engineered crops, and transgenic pigs. At other times and in other places—in other cultures—what we might describe as tools, or even technology, were, or are, considered, like an artist's paintbrush, not a tool that could be used for whatever purpose someone might conceive of, but an extension of the artist's hand itself. The tool, the technology, does not exist apart from its agent or outside of a specific culture. The process of seed selection in subsistence agriculture, for example, is not a tool or technology for *improvement* of a crop but an integral aspect of the relationship between the farmer, the plant, and the land. In what is now assumed to be the dominant and *correct* culture of food production, however, the seed has been reduced to a commodity produced or engineered without an intimate relationship with any specific context by a corporation that regards the seed as primarily a means of producing wealth. The seed is reduced to a technology whose value is no longer intrinsic in its ability to provide sustenance for life, but derived exclusively from its value as a commercial product, quite apart from its possible, or essential, sustenance value.

In the practice of biotechnology, not only is there no formal acknowledgment of responsibility for the technological products or processes (hence the problem of insurance, or lack thereof), but now there is a concerted effort to formally deny responsibility by shifting the burden of proof (onus) onto the regulatory agencies—that is, the public—or those accused of infringing on the "rights" of corporation in the case of patents. In other words, there is underway an attempt to make the public—whether the regulatory agency or an individual who is directly affected—responsible for proving harm or lack of safety, rather than the protagonist or applicant for approval being required to prove safety of biotechnology product or process. The official explanation for this reversal is that we—the government regulators, the public—must not unduly obstruct the economic enterprise of getting new products to market. (The patenting of life-forms follows a similar logic.)[11]

In the same way, the peasant's bullock is not just a draft animal or a beast of burden. It is a provider of milk, fertilizer, and next year's meat, as well as a companion. The modern dairy farm, in contrast, consists of a herd of genetically near-identical cows whose value is measured electronically in terms of the kilos of milk given over the course of a lactation. Then cow herself is probably "recognized" as a number on an ear tag that can be read electronically, though even in the occasional "modern" North American dairy herd, one can sometimes find a cow of sixteen or even twenty years enjoying her privileged retirement, an expression of the farmer's respect and appreciation for her life's work.

The environment of all technology is a human community (not *the* human community), and the technology of a community has always been an extension of its capabilities and an expression of its culture. The technology of one community might, however, be the junk of another, or simply a mystery to be gawked at. This is as true today as it ever was. Even the pretensions and delusions of the purveyors of globalization are not sufficient to constitute a universal monoculture, as attested to by the apparent metaphysical requirement to impose, by legal or extralegal means, such as the World Trade Organization, the monoculture of "globalization" and the "technology" of the imperial industrial states.

All technologies, biotechnologies included, are indeed cultural artifacts, designed and constructed as expressions of specific cultures at particular times and in particular places. The people who construct these technologies follow their own agenda. The technology they come up with may be a simple tool to help them get on with the task at hand, or it may be an end in itself, what we might now refer to as a product for the market. But it may still be considered a tool, such as a specialized kitchen gadget, an internal combustion engine or a fuel cell designed to power a means of locomotion. It is quite appropriate, then, to describe agricultural biotechnology as "a powerful tool," as the promoters of biotechnology are fond of doing. The genetically engineered characteristic that enables a plant to tolerate a particular, otherwise deadly, herbicide becomes, supposedly, a tool for increased production, easier management, greater efficiency, or some other extrinsic "value." Similarly, the toxin engineered into corn or soy to kill a specific predator is merely a tool. The genetic constructs of biotechnology are denatured to become just another tool in the farmer's toolbox. It's a clever way to avoid—and to obscure—the social and ethical issues that may be asked of and by such tools.

This is, however, a particular and, to many, a peculiar way of regarding organisms or their parts. It is a utilitarian approach to life-forms that makes it difficult, if not impossible, to discuss the ethics of biotechnology. Questions of right and wrong, good and bad, are not applicable to what is defined or described as a technology if technologies are simply universal gadgets, value-free tools, without context of culture or environment.

Many of the ethical issues that one would think would be raised by agricultural biotechnology—issues such as environmental loading and risk, commodification of life, control of Nature, or elite benefit versus social good—are really questions about the larger program and the values of its context, about the culture that has developed such tools as capitalism, reductionist science, Enlightenment rationality, and progress. Other cultural strands that contribute to the fabric of biotechnology, as suggested earlier, are monoculture, technological determinism, Nature as alien and enemy, and individualism. Obviously, any such listing is both arbitrary and limited, and the ordering of such characteristics—the question of which is a subset of which—is certainly open to debate.

Each of the issues or characteristics listed above are remarkable in that they are not about an organism or our relationship to it per se—plant or animal, single-celled or multicellular—but about the character of the human context of the organism, that is, the human agency that created the novel product and the world it constructs and inhabits.

Identifying and describing (or characterizing) the context, the culture that has produced genetic engineering, is not a simple task. States and nations (political jurisdictions and societies) such as Canada and the United States are no longer the ethnic or religious monocultures that the ideology of science and technology assumes them to be. They probably never were.

We are, however, neither trained nor encouraged to look into or through the technology to discern the human agency and intent behind or within it, for good or for evil. We assume that the only ethical issues to be dealt with are at the interpersonal level: how we treat one another as we pursue a common, agreed-upon purpose, or how we are to interact with the technology. Cultural differences are regarded not as fundamentally different outlooks on life, or even differences of faith, but as superficial folkways and colorful but insignificant differences in dress, food habits, and even in religion. Of course we all want to own a car, use a cell phone, eat "improved" industrial food ("more nutritious, better tasting"), and have "perfect" babies.

The result is that biotechnology is put forward as just another technology in the tradition of Western culture that treats all technology, like science, as ethically and morally neutral, or simply beyond ethical or moral judgment. It's all a matter of who uses the technology and for what purposes. The end justifies the means. Development and use of the atom bomb, the deterrence doctrine of the Cold War years, and now the nuclear shield are prime examples. The current development of biological warfare agents proceeds on the basis of the same rationale, but brings us closer to the issue of biotechnology in agriculture. The intentional use of agrotoxins to wage a wholesale war on certain crops (coca), against the will of the inhabitants of the land and life being violated, reflects this same logic of ends justifying means. But then the organisms inhabiting the lands of industrial agriculture have never been consulted about their treatment, either. They have never been party to an ethic of informed consent, to say nothing of an ethic of respect and appreciation for boundaries.

In the case of agricultural biotechnology, ethics has been reduced to an enteric coating on the bitter pill being forced down the public throat.

Chapter Four

When Transgenes Wander, Should We Worry?

Norman C. Ellstrand

When Maarten Chrispeels invited me to write an editorial explaining the science behind one of the environmental risks of biotechnology for the readership of Plant Physiology, I was pleased to do so. The editorial was one in a series that examined issues associated with transgenic plants. The readership of that journal is comprised of a variety of plant scientists, including many lab-based ones who might have little access to the population genetic or ecological literature. After publishing the article, I received a surprising amount of positive feedback. Therefore, if I had to do it all again, there is little that I would change. However, a few details are worthy of updating. First, although I made it clear that weed beets have created substantial problems for Europe's sugar industry, I should have explained that the industry has been suffering problems on other fronts as well. Second, a scientific article has recently been published that describes the evolution of multiple herbicide resistance in Canadian canola.[1] Third, the reader should be aware that a subset of plant genetic engineers are aware of the potential problems from gene flow—and that some suggestions have been made on how to contain transgenes.[2] To the best of my knowledge, however, these suggestions remain theoretical, even though containing transgenes would also serve to prevent the unintentional release of intellectual property, a great concern for the industry as a whole.[3] I am grateful to have been invited by the editor of this volume to reprint the article so that it can enjoy a new readership alongside the works of so many scientists whom I hold in high regard.

It is hard to ignore the ongoing, often emotional, public discussion of the impacts of the products of crop biotechnology. At one extreme of the hype is self-righteous panic, and at the other is smug optimism. While the controversy plays out in the press, dozens of scientific workshops, symposia, and other meetings have been held to take a hard and thoughtful look at potential risks of transgenic crops. Overshadowed by the loud and contentious voices, a set of straightforward, scientifically based concerns have evolved, dictating a cautious approach for creating the best choices for agriculture's future.

Plant ecologists and population geneticists have looked to problems associated with traditionally improved crops to anticipate possible risks of transgenic crops. Those that have been most widely discussed are: (a) crop-to-wild hybridization resulting in the evolution of increased weediness in wild relatives, (b) evolution of pests that are resistant to new strategies for their control, and (c) the impacts on nontarget species in associated ecosystems (such as the unintentional poisoning of beneficial insects).[4]

Exploring each of these in detail would take a book, and such books exist.[5,6] However, let us consider the questions that have dominated my research over the last decade to examine how concerns regarding engineered crops have evolved. Those questions are: How likely is it that transgenes will move into and establish in natural populations? And if transgenes do move into wild populations, is there any cause for concern? It turns out that experience and experiments with traditional crops provide a tremendous amount of information for answering these questions.

The possibility of transgene flow from engineered crops to their wild relatives with undesirable consequences was independently recognized by several scientists.[7,8,9] Among the first to publish the idea were two Calgene scientists, writing: "The sexual transfer of genes to weedy species to create a more persistent weed is probably the greatest environmental risk of planting a new variety of crop species."[10] The movement of unwanted crop genes into the environment may pose more of a management dilemma than unwanted chemicals. A single molecule of DDT [1,1,1,-trichloro-2,2-bis(p-chlorophenyl)ethane] remains a single molecule or degrades, but a single crop allele has the opportunity to multiply itself repeatedly through reproduction, which can frustrate attempts at containment.

In the early 1990s, the general view was that hybridization between crops and their wild relatives occurred infrequently, even when they were growing in close proximity. This view was supported by the belief that the discrete evolutionary pathways of domesticated crops and their wild relatives would lead to increased reproductive isolation and was supported by challenges breeders sometimes have in obtaining crop-wild hybrids. Thus, my research group set out to measure spontaneous hybridization between wild radish (*Raphanus sativus*), an important California weed, and cultivated radish (the same species), an important California crop.[11] We grew the crop as if we were multiplying commercial seed and surrounded it with stands of weeds at varying distances. When the plants flowered, pollinators did their job. We harvested seeds from the weeds for progeny testing. We exploited an allozyme allele (Lap-6) that was present in the crop and absent in the weed to detect hybrids in the progeny of the weed. We found that every weed seed analyzed at the shortest distance (1 m) was sired by the crop and that a low level of hybridization was detected at the greatest distance (1 km). It was clear, at least in this system, that crop alleles could enter natural populations.

But could they persist? The general view at that time was that hybrids of crops and weeds would always be handicapped by crop characteristics that are agronomically favorable, but a detriment in the wild. We tested that view by comparing the fitness of the hybrids created in our first experiment with their non-hybrid siblings.[12] We grew them side by side under field conditions. The hybrids exhibited the huge swollen root characteristic of the crop; the pure wild plants did not. The two groups did not differ significantly in germination, survival, or ability for their pollen to sire seed. However, the hybrids set about 15% more seed than the wild plants. In this system, hybrid vigor would accelerate the spread crop alleles in a natural population.

When I took these results on the road, I was challenged by those who questioned the generality of the results. Isn't radish probably an exception? Radish is outcrossing and insect pollinated. Its wild relative is the same species. What about a more important crop? What about a more important weed? We decided to address all of those criticisms with a new system. Sorghum (*Sorghum bicolor*) is one of the world's most important crops. Johnsongrass (*Sorghum halepense*) is one of the world's worst weeds. The two are distinct species, even differing in chromosome number, and sorghum is largely selfing and wind pollinated. Sorghum was about as different from radish as you could get.

We conducted experiments with sorghum paralleling those with radish. We found that sorghum and johnsongrass spontaneously hybridize, although at rates lower than the radish system, and detected crop alleles in seed set by wild plants growing 100 m from the crop.[13] The fitness of the hybrids was not significantly different from their wild siblings.[14] The results from our sorghum-johnsongrass experiments were qualitatively the same as those from our cultivated radish–wild radish experiments. Other labs have conducted similar experiments on crops such as sunflower (*Helianthus annuus*), rice (*Oryza sativa*), canola (*Brassica napus*), and pearl millet (*Pennisetum glaucum*).[15] In addition, descriptive studies have repeatedly found crop-specific alleles in wild relatives when the two grow in proximity.[16] The data from such experiments and descriptive studies provide ample evidence that spontaneous hybridization with wild relatives appears to be a general feature of most of the world's important crops, from raspberries (*Rubus idaeus*) to mushrooms (*Agaricus bisporus*).[17]

When I gave seminars on the results of these experiments, I was met by a new question: "If gene flow from crops to their wild relatives was a problem, wouldn't it already have occurred in traditional systems?" A good question. I conducted a thorough literature review to find out what was known about the consequences of natural hybridization between the world's most important crops and their wild relatives.

Crop-to-weed gene flow has created hardship through the appearance of new or more difficult weeds. Hybridization with wild relatives has been implicated in the evolution of more aggressive weeds for seven of the world's 13 most important crops.[18] It is notable that hybridization between sea beet (*Beta vulgaris* subsp. *maritime*) and sugar beet (*B. vulgaris* subsp. *vulgaris*) has resulted in a new weed that has devastated Europe's sugar production.[19]

Crop-to-wild gene flow can create another problem. Hybridization between a common species and a rare one can, under the appropriate conditions, send the rare species to extinction in a few generations.[20,21,22] There are several cases in which hybridization between a crop and its wild relatives has increased the extinction risk for the wild taxon.[23] The role of hybridization in the extinction of a wild subspecies of rice has been especially well documented.[24] It is clear that gene flow from crops to wild relatives has, on occasion, had undesirable consequences.

Are transgenic crops likely to be different from traditionally improved crops? No, and that is not necessarily good news. It is clear that the probability of problems due to gene flow from any individual cultivar is extremely low, but when those problems are realized, they can be doozies. Whether transgenic crops are more or less likely to create gene flow problems will depend in part on their phenotypes. The majority of the "first generation" transgenic crops have phenotypes that are apt to give a weed a fitness boost, such as herbicide resistance or pest resistance. Although a fitness boost in itself may not lead to increased weediness, scientists engineering crops with such phenotypes should be mindful that those phenotypes might have unwanted effects in natural populations. In fact, I am aware of at least three cases in which scientists decided not to engineer certain traits into certain crops because of such concerns.

The crops most likely to increase extinction risk by gene flow are those that are planted in new locations that bring them into the vicinity of wild relatives, thereby increasing the hybridization rate because of proximity. For example, one can imagine a new variety that has increased salinity tolerance that can now be planted within the range of an endangered relative. It is clear that those scientists creating and releasing new crops, transgenic or otherwise, can use the possibility of gene flow to make choices about how to create the best possible products.

It is interesting that little has been written regarding the possible downsides of within-crop gene flow involving transgenic plants. Yet a couple of recent incidents suggest that crop-to-crop gene flow may result in greater risks than crop-to-wild gene flow. The first is a report of triple herbicide resistance in canola in Alberta, Canada.[25] Volunteer canola plants were found to be resistant to the herbicides Roundup (Monsanto, St. Louis), Liberty (Aventis CropScience, Research Triangle Park, NC), and Pursuit (BASF, Research Triangle Park, NC). It is clear that two different hybridization events were necessary to account for these genotypes. It is interesting that the alleles for resistance to Roundup and Liberty are transgenes, but the allele for Pursuit resistance is the result of mutation breeding. Although these volunteers can be managed with other herbicides, this report is significant because, if correct, it illustrates that gene flow into wild plants is not the only avenue for the evolution of plants that is increasingly difficult to manage.

The second incident is a report of the StarLink Cry9C allele (the one creating the fuss in Taco Bell's taco shells) appearing in a variety of supposedly nonengineered corn.[26] Although unintentional mixing of seeds during transport or storage may explain the contamination of the traditional variety, inter-varietal crossing between seed production fields could be just as likely. This news is significant because, if correct, it illustrates how easy it is to lose track of transgenes. Without careful checking, there are plenty of opportunities for them to move from variety to variety.

The field release of "third generation" transgenic crops that are grown to produce pharmaceutical and other industrial biochemicals will pose special challenges for containment if we do not want those chemicals appearing in the human food supply.

The products of plant improvement are not absolutely safe, and we cannot expect transgenic crops to be absolutely safe either. Recognition of that fact suggests that creating something just because we are now able to do so is an inadequate reason for embracing a new technology. If we have advanced tools for creating novel agricultural products, we should use the advanced knowledge from ecology and population genetics as well as social sciences and humanities to make mindful choices about to how to create the products that are best for humans and our environment.

Acknowledgments

This article was written while I was receiving support from the U.S. Department of Agriculture (grant no. 00-33120-9801). I thank Tracy Kahn for her thoughtful comments on an earlier draft of the manuscript and Maarten Chrispeels for his encouragement and patience.

This chapter appears with the permission of ©Plant Physiology, 2001. Ellstrand, N.C. "When transgenes wander, should we worry?" *Plant Physiology*, 2001; 125:1543–1545.

Chapter Five

Patents, Plants, and People:
The Need for a New Ethical Paradigm

Lori B. Andrews

In the beginning, there was life. Whether one imagines a Garden of Eden or a sea of slimy amoebas, life-forms have enlivened our planet for eons. In the future, though, there may be only commerce. *Homo sapiens* may be replaced by *Homo Glaxo Wellcomus* or even Merck Man.[1] The only soy available may be Monsanto soy. Certain species may continue to exist only if they find a corporate sponsor.

What are the implications of the corporatization of life-forms? Application of a business model to living entities poses the following major dangers.

- *In business, products are promoted, often in ways that lead to irrational decisions.* The downsides may be relatively small if it means buying designer sneakers or a car that can reach speeds the driver will never use, but it might not be the best way to create food or pets or children. Promoting a genetically modified food so that it has an enormous market share may, by narrowing the range of foods a particular person eats, lead to deficiencies in diet.[2]
- *Marketing may involve "scare" tactics.* Witness what is going on with for-profit companies' pressure on parents to "bank" their childrens' cord blood commercially as "biological insurance," even though there is only a minuscule (1 in 20,000) chance the child will later

need the cord blood and it would be better for all children if cord blood were available through public nonprofit banks. Or look at the pressure placed on parents of short/normal children to use bioengineered human growth hormone. One pediatric endocrinologist coerces parents, asking what they will tell their short son when he grows up and learns he could have been 5 foot 10 inches tall.[3] Similar pressures may be exerted to convince people to eat foods that have been engineered to contain certain drugs or vitamins, despite the potential risk of overdose.

- *In business, short-term gains are emphasized and long-term risks are often ignored.* Environmental pollution may be replaced by genetic pollution—risks to future generations due to genetic modifications of food products or people. For example, germ line genetic interventions (heritable genetic modifications of sperm, eggs, or embryos) create changes that are inherited by subsequent generations and may increase cancer risks in the next generation.[4] Ingestion of genetically modified food with antibiotic markers may increase antibiotic resistance in the future.[5]

- *Corporate monopolies may grow.* Already, the annual profits of some companies involved in genetic modification of life-forms are greater than the gross national products (GNPs) of many individual countries. For example, in 1998, Monsanto made net sales of $4 billion in agricultural products,[6] which is more than the GNP of many African nations, Cambodia, or Nicaragua.[7]

- *Diversity of life-forms may diminish.* Of the soybean acreage in the United States, more than one-half is genetically modified—almost all of which is Monsanto's Roundup Ready.[8] The loss in food crop diversity is also exemplified by seed companies whose holdings have shrunk. This is due to the fact that a few companies control a significant portion of the global food market, and as they focus on the bottom line of profitability, they are turning to biotechnology and narrow product lines as a way of protecting their investments.[9] Currently, about 31 percent of the global commercial seed market is dominated by ten seed companies, and five vegetable seed companies control 75 percent of the global vegetable seed market.[10] Between 1993 and 1995, nine hundred and fifty varieties of vegetables became extinct and nearly four thousand remaining varieties are endangered.[11] The history and practices of Seminis, the world's largest vegetable seed corporation, demonstrates how corporate mergers are contributing to the shrinking seed market.[12] Seminis is a subsidiary of Mexico-based Savia, which controls 40 percent of the U.S. vegetable seed market. Savia built the company by acquiring about a dozen seed companies, including Asgrow, Petoseed, and Royal Sluis, and today has a presence in about 120 countries. Seminis plays a large role in the development of genetically engineered vegetables; it currently owns at least seventy-nine patents and has others pending, including some relating to beans, bean sprouts, corn, and cauliflower.

On June 28, 2000, Seminis declared that it planned to eliminate 25 percent, or two thousand, of its varieties as part of its "global restructuring and optimization plan." Under this market-driven approach, Seminis prefers plants that are hybrids, because farmers cannot replant them for the next season and must purchase seeds annually.[13] Yet when Seminis decides to stop using seeds that it views as obsolete and unprofitable, it fails to make that information public. To the extent that measures could be taken by a number of groups concerned about disappearing varieties, such as U.S.-based Seed Savers Exchange (SSE), they are denied access to important information. Once the seed is acquired by a company, that enterprise can decide not to use it, at which point it becomes part of the company's private gene bank and not available to the public.

• *People may be treated as products.* As an increasing number of other life-forms are "enhanced" through genetic manipulation, the view of biology as a set of Legos or Tinker Toys will inevitably lead to attempts to genetically engineer people. In the future, human embryos might be genetically manipulated and then patented. Already, there has been a patent application for a process to genetically engineer mammals to produce pharmaceutical products in their milk. The application asks for the rights to patent genetically engineered *human women* as well. Brian Lucas, the British patent attorney for Baylor, said that although the focus of the current technology was cows, the desire to cover women was put in because "someone, somewhere may decide that humans are patentable" and Baylor wanted to protect its intellectual property if that happened.[14]

Corporatization of Food and Animal Production
Most people do not realize the major transformation that is occurring in our lives via the issuing of patents on the elements of life (such as genes) and on genetically engineered life-forms. People who do think about it tend to brush it off with comments such as, "Well, we've bred

animals and plants for centuries." But something dramatically different is now occurring. Certain scientists and their corporate sponsors are claiming, under patent law, to own ideas, not just inventions. They are also claiming to have "invented" plants and animals, merely because they have added a single gene or a few genes to a complex, multifaceted organism. They are also claiming to own not only the living entity, but its offspring as well. Thus, if I buy a rabbit that has been genetically engineered to glow green, and that rabbit escapes and reproduces with a wild rabbit, I will owe the patent holder licensing fees for each of the bunnies created.[15] This is because a patent holder can control for twenty years the "making" of any such green rabbit.

As this profound social transformation is taking place, a series of questions need to be addressed:

- Should genes be owned?
- Should life-forms be patented?
- What respect is due to the human or cultural sources of genes?
- What weight should be given to the existing biological limitations embodied in species barriers to reproduction?

Imagining a Past Full of Patents

One way to judge the appropriateness of patenting in the biological realm is to imagine what our life would have been like if these patent procedures had been applied in the past. Consider the first brave person to drink milk from a cow or to eat a cooked chicken. What if he had patented that process? (At the time, it was new, non-obvious, and useful, thus meeting the basic requirements for a patent.) That "inventor" would have been able to control the technology for the next twenty years. Not only could he have charged whatever he wanted to anyone who drank milk or cooked chicken, he could have forbidden people he didn't like from drinking or eating these substances.

And what if, a generation ago, geneticists had patented their findings? Long before the advent of the Human Genome Project, genetics researchers noted the connection between an individual's genetic makeup and diseases or disorders.[16] In 1959, Jerome Lejeune discovered the association of Down syndrome with chromosome 21. Yet he did not patent the observation.[17] If he had, then for two decades, he would have controlled any genetic test or genetic treatment associated with the disorder. The prenatal test for the disorder might have become prohibitively expensive, and research collaboration might have been deterred.

Moreover, Dr. Lejeune happens to believe that life begins at conception.[18] If he had gained a patent, he would have been able to prohibit anyone from doing prenatal testing for Down syndrome—in order to deter abortion. Mark Skolnick, who patented the BRCA1 gene for diagnosis of inheritable breast cancer, says he will block its use in prenatal diagnosis.[19] Whether or not one thinks it is appropriate to test prenatally for Down syndrome or breast cancer, I don't think anyone would say it is reasonable for that profound ethical question to be decided exclusively at the whim of the patent holder. Skolnick, just because he holds the patent, is making *illegal* an otherwise legal act.

A similar issue arises in agriculture biotechnology. The Plant Varieties Act permits agricultural research involving plants. In contrast, the holder of a patent on a genetically modified plant can prohibit such research. The Plant Variety Protection Act allows farmers to save seeds from a proprietary crop. But patented genetic technologies can be used to produce a toxin late in seed development to create sterile seeds so the farmer cannot plant the retained seeds the following season. Law professor Dan Burk refers to this possibility as "Lex Genetica," since the principles of existing law can be evaded by the owners of a patented genetic technology.[20]

Should Genes Be Owned and Life-Forms Patented?

At the heart of the new commercialization of life-forms is the patent system. The recombinant DNA technology invented by Herbert Boyer and Stanley Cohen and patented in 1976 seeded a vast new biotech industry that is based on treating genes, other biological products, and entire life-forms as raw materials for commercial products.[21] In a 1980 case, *Diamond v. Chakrabarty,* a researcher sought a patent on a living bacterium that was genetically engineered for use in cleaning up oil spills.[22] The U.S. Patent and Trademark Office denied the patent application on the grounds that living organisms were not patentable. However, the U.S. Supreme Court, reviewing the matter, observed that the organism was not an unpatentable "product of nature" since such genetically manipulated bacteria are not found in nature. The test for whether something was patentable, said the Court, was whether the invention was the result of human intervention.[23] "Anything under the sun that is made by man" could be patented, said the Court[24]—so long as it met other patent law standards of being novel,[25] non-obvious,[26] and useful.[27]

Most of the legal documents filed in the *Chakrabarty* case focused only on the bacterium. Just one amicus brief raised concerns about how the Court's decision about the patenting of the bacterium might affect people. "One day," cautioned the brief of the Peoples Business Coalition, "it will be possible to convert higher organisms, including human beings, into 'industrial products' just as microorganisms are being so engineered today."[28]

The *Chakrabarty* decision gave molecular biologists the assurance they would own any life-forms they invented by combining genes in ways that did not occur in nature. But when the Human Genome Project proposed to spend $3 billion of taxpayer money identifying the more than thirty thousand human genes, most biological scientists had no expectation they would actually own the genes they studied. In fact, the idea at that time seemed absurd. The federal patent statutes assure that "whoever invents or discovers any new and useful process, machine, manufacture, or composition of matter, or any new and useful improvement thereof, may obtain a patent."[29] But patent law prohibits patenting a product of nature.[30] It also prohibits patenting a formula, like $E=mc^2$. Genes are both.

Because gene patenting seemed inconceivable at that time, Nobel laureate Walter Gilbert announced a scheme to copyright DNA, just as you would a book.[31] His plot was to own a gene sequence, such as CATTAGTA . . . , and charge a fee for access to information about which sequences corresponded with which genes. The idea went nowhere. The idea of an individual having dominion over the common language of genes seemed overreaching and ludicrous.

Key researchers in the field at the time—C. Thomas Caskey, then at Baylor University and Leroy Hood, then at Cal Tech—warned about the risks of patenting. They noted that if scientists were allowed to patent genes and reap financial rewards by having exclusive rights to any diagnostic or treatment technologies developed with the gene they found, researchers would be less likely to share copies of the genes they discovered or even to share information about those genes.[32]

Entrepreneurs seeking patents on genes, though, pointed to precedents allowing patents on naturally occurring substances such as vitamin A that were isolated and purified. They claimed they had "isolated" DNA from the human body. But the vitamin A patent holders sold a product—a dietary supplement. Patent holders on genes claim dominion over information—the genetic sequence itself. And they can extract a licensing fee from anyone who tries to diagnose whether a person has that particular genetic sequence in his or her body.[33]

Today, researchers are patenting not only whole genes but small subparts of genes and mutations of genes.[34] Although there are only an estimated 30,000 genes, patent applications had been filed on over 126,672 genes or partial gene sequences.[35] As smaller and smaller sections of genes are patented, licensing becomes more of a constraint. Although Myriad has a patent on the breast cancer gene, other researchers have patented additional mutations. This means that any institution offering a breast cancer gene test (or doing research on gene therapy for the disease) must first obtain licenses from all patent holders. Myriad's test itself costs $2,580. Now imagine the price when the hundreds of additional mutations that are known to exist in the breast cancer gene are also patented. Even if each patent holder only asked for $10 each, the cost for the resulting test that checked for all mutations would go through the roof.

Within the past two years, access to genetic tests has decreased substantially as gene patents have been granted. One in four laboratories has stopped performing certain genetic tests because of patent restrictions or excessive costs.[36] The Ashkenazi Jewish community in Brooklyn used to routinely undergo premarital tests for carrier status for the genetic disorder Canavan disease, which causes a fatal brain disease in childhood. The screening program was stopped for a while when the patent holder demanded a high licensing fee,[37] and now may have to be dropped altogether.[38]

Patenting genes—whether of humans, animals, plants, or other life-forms—also impedes important research. The sharing of research materials is decreasing.[39] Scientists with access to biological materials or genes are now less likely to give samples of those materials to other researchers. This is true even though replication techniques can create millions of copies of the genes or cells and thus sharing does not diminish the first scientist's ability to carry out research.

Gene patents can deter innovation in biomedical research. "A proliferation of intellectual property rights upstream may be stifling life-saving innovations further downstream in the course of research and product development," wrote Michigan law professors Michael A. Heller and Rebecca Eisenberg in *Science*.[40] They likened the situation in genetics to that of postsocialist economies. The expectation in Eastern Europe was that private stores would be loaded with goods once a free market was established. But the stores remained bare—while street vendors flourished. The reason—no individual could set up shop without collecting rights from workers' collectives, privatization agencies, and local, regional, or federal governments. Similarly, with genetics, "privatization can go astray when too many owners hold rights in previous discoveries that constitute obstacles to future research."[41] Since there are more than a hundred patents, for example, related to the adrenalin receptor, a researcher whose work is related to that site faces a daunting bargaining procedure.[42]

In the past, patents were sought for things—such as a mousetrap. But now, patents are being sought for broad concepts, equivalent to attempting to patent the idea of trapping mice,[43] allowing the holder of a patent on a microscope to gain rights to anything discovered by looking through the lens.

Yet patents on broad concepts affecting whole fields seem far removed from the idea of the patent system—which was to encourage specific inventions. The absurdity of the new patenting mentality is underscored by a law firm's recent suggestion that professional athletes should patent unique moves.[44] But if someone patents a unique style of dunking, no other player would be able to use it. Similarly, patenting genes and life-forms gives the "owner" unprecedented and unwarranted control.

What Respect is Due the Human or Cultural Sources of Genes?

The researcher who patents a gene, cell line, plant, or animal is claiming ownership over a life-form that in a very real sense belongs to someone else, such as the human source of a gene or the indigenous culture that created a plant with special properties. The court case *Moore v. Regents of the University of California* dealt with a doctor who patented a patient's cell line, apparently without the patient's knowledge or consent. The patient, John Moore, at first reacted with disbelief. Then, as he thought more about what had happened, he felt "violated for dollars," "invaded," "raped."[45] His body had been appropriated without his knowledge or consent. He sued his doctor for theft. The California Supreme Court responded by saying an individual could not have a property interest in his cell line, but a doctor or biotech company could! The court did give Moore a right to sue his doctor for lack of informed consent and for breach of what is known as the doctor's "fiduciary duty"—his responsibility to put the patient's interest first. They held that a physician must tell his patient if he has a personal interest unrelated to the patient's health, whether research or economic, that might affect his judgment.

A dissenting justice was outraged. He pointed out that the doctor had not created Moore's cell line. The doctor, he said, "could not have extracted the Mo cell line out of thin air."[46] Justice Mosk pointed out that John Moore's contribution was "absolutely crucial: . . . but for the cells of Moore's body taken by defendant there would have been no Mo cell line."[47] Yet the rights of the individual and cultural sources of life-forms are often violated in the course of commercializing life.

John Moore's situation was not unique. Over 30 percent of patents filed by medical schools are for products based on patients' tissues or genes.[48] Researchers are increasingly recognizing that people from isolated populations in the Third World may have unique body tissue. In March 1995, researchers from the National Institutes of Health (NIH) obtained a patent on the DNA of a man from New Guinea, whose genes protect him from leukemia. As scientists from the NIH and various biotechnology companies search the globe for DNA that presents promise for research or commercialization, protests arise, pointing out that such intrusions violate personal and cultural integrity. The targets of bioprospecting are treated not as people but as "genetic isolates"—in the words of the scientific effort known as the Human Genome Diversity Project (HGDP). As one representative of an indigenous group opined, "You've taken our land, our language, our culture, and even our children. Are you now saying you want to take part of our bodies as well?"

The same argument could apply to the control of genes derived from indigenous plants. The recent trend of patenting genes from people mirrors the growing trend of multinational companies in patenting plant germ plasm that had been cultivated by indigenous populations. In Latin America, for example, centuries of cultivation and breeding by local farmers led to a strain of cotton that was colored. The U.S. Department of Agriculture brought back seeds—and an American scientist, Sally Fox, patented the colored cotton, listing herself as the "inventor."[49] The local farmers apparently received no portion of the credit or the profits.[50]

The neem tree had been used for centuries in India because of medicinal and pesticidal properties. In the mid-1990s, though, W. R. Grace and other U.S. firms filed patents on processes involving, and products from, neem seeds. In April 2000, the European Patent Office (EPO) revoked the 1995 patent for a process to extract oil from neem trees to be used as a pesticide. The patent was revoked based on the EPO's finding that it lacked invention and novelty. The EPO found that farmers in India had previously used this process.[51] In fact, India has seen so many of its indigenously developed life-forms patented by American researchers that Indians have created a computerized database that is searchable by the U.S. Patent Office to try to establish, by prior art, that these crops should not be patentable.[52]

Proposals have been made to share scientific credit or some modest patent royalties with the people or cultures whose biological essences are being patented. But to many indigenous groups, the idea of commercializing life itself is totally inappropriate. They do not want to share in the patents; they want to abolish them.

When John Moore traveled to India to protest the patenting of human genes, hundreds of farmers turned out to meet him. Many cried when he told his story. Ultimately, protests against the audacity of the U.S. government in patenting the genes of foreign citizens were sufficiently widespread that in 1995 patents on the genes of people from South America, the Solomon Islands, and Papua, New Guinea, were withdrawn.[53]

In 1996, a group of indigenous leaders picketed a meeting of scientists and ethicists in Montreal where a session was held to discuss the Human Genome Diversity Project. Their signs revealed their concerns about the project: "No Human $ell lines," "Preserve our culture, not tissue culture," and "*Alto al Projeto Vampiro*," which depicted a vampire with blood dripping off of its fangs. The project, claimed the protestors, was violating the cultural beliefs of indigenous people and exploiting them economically, as the profits from their bodies would—like other natural resources—be used for the benefit of people in developed nations.[54]

What Weight Should Be Given to the Existing Biological Limitations Embodied in Species Barriers to Reproduction?

The manipulation of life-forms by introducing genes from one species into another is increasingly occurring in the agricultural biotech sector and the medical research sector. Attempts have been made to add fish genes to strawberries to make it less likely the plants will freeze.[55]

Human cancer cells have been added to mice.[56] A genetically altered potato is classified by the EPA as a pesticide.[57] In the United States, more than forty-five hundred types of genetically modified plants have been tested.[58]

Genetic manipulation across species barriers will likely move rapidly into other sectors of society. Clinical medicine, including fertility services, will begin to offer to "upgrade" people by adding genes from other species. In 1999, California artists Tran T. Kim Trang and Karl Mihail leased a store in a trendy shopping area of Pasadena and purported to sell genetic traits. The exhibit was so realistic that people plunked down their credit cards to order genetic modifications on themselves (including one man who wanted the longevity of a cockroach).[59]

All this suggests we are moving toward a new eugenics, where all genetic units, no matter what their source, seem interchangeable. In a Louis Harris poll sponsored by the March of Dimes, 42 percent of potential parents said they would use genetic engineering on their children to make them smarter; 43 percent, to upgrade them physically.[60] Species boundaries are no longer relevant. In the future, people might attempt to give their children the running speed of a cheetah or the night vision of a bat.

The value of human life will diminish by treating children increasingly like other consumer goods, such as cars to be ordered with a choice of extras. Already, some parents have brought product liability–type suits against infertility clinics. In one, the mother alleged her children would have been more attractive if a different sperm donor had been used.[61]

The Troubling Impact of Genetic Interventions

Although profound genetic interventions are occurring in a variety of realms, little attention is being paid to the physical, psychological, or social risks created by the crossing of species barriers. Yet both short- and long-term consequences can be profound.

Genetic engineering across species in agriculture may cause allergic reactions in people who eat the resulting foods. When a gene from the Brazil nut was transferred to soybeans, people allergic to nuts were allergic to the genetically altered soybeans.[62] Prohevein, one of the antigens from the rubber tree *Havea brasilensis* that causes the latex allergy in some individuals, has been engineered into tomatoes for its fungastatic properties. Allergist David Freed, writing in the *British Medical Journal* predicted that "we can expect an outbreak of tomato allergy in the near future among latex sensitive people."[63] In Great Britain, there has been a 50 percent increase in soybean allergies,[64] some of which may be attributable to the gene-altered products contained therein.

In the past five years, genetically modified foods have surreptitiously entered the diets of most Americans. Yet, because producers and sellers are not required to indicate which foods have been genetically modified, people with potential allergies cannot protect themselves. Moreover, genetically modified foods may be nutritionally inferior. Roundup Ready soybeans may have reduced levels of phytoestrogens, which lower cholesterol.[65] Other life-forms may be harmed by the use of genetically modified plants. Some farmers using Roundup Ready soybeans have used two to five times more herbicide than in the past.[66] Studies indicate that crops expressing the Bt pesticide may kill beneficial insects.[67] When plants are genetically engineered to resist viruses, the viruses may mutate into more dangerous forms.[68]

Even some of the genetic technologies meant to benefit health could backfire. The July 2000 issue of the *Journal of Infectious Diseases* reported a study where potatoes were genetically engineered with a vaccine against the Norwalk virus, triggering an immune response to *E. coli.*[69] It was touted as a cheap, effective way to vaccinate people, especially in the Third World. But putting a vaccine in food eliminates the ability to control dosage. People who eat a lot of potatoes could get a toxic overdose. Eating too few potatoes might cause disease outbreaks among people presumed to be immune.[70]

Farming is being profoundly transformed.[71] Certain companies are gaining a monopoly control over genetic resources. Breeders no longer have free access to germ plasm for developing new varieties of plants and animals. Consumers will likely pay more for food, medicine, and other biotech products.

Life patents threaten fundamental values. Human rights will erode as people's biological materials—and, ultimately, people themselves—become the exclusive property of patent holders. And instead of respect for life based on ethical and religious values, society's relationships to nature will be restructured to one that is reductionistic and commercial.

The "Costs" of Commercialization

The patenting of life-forms represents enormous hubris. Scientists and companies are claiming to have "invented" products of nature. It is a telling paradox: On the one hand, the companies have persuaded the Food and Drug Administration and the U.S. Department of Agriculture that they should not have to label genetically modified food because the inserted genes are products of nature (no different from their conventionally grown counterparts). Yet, those same companies are telling the patent office that genes are not products of nature and that the companies should own the genes as their inventions.

Genes *are* products of nature, and thus should not be patented. But just because they are "natural" does not mean that they are safe when transferred to other species. Nor does it mean that the process of turning genes and life-forms into raw materials and commercial products can be undertaken without a profound diminution in societal values.

The genetic modification and patenting of plants, under corporate auspices, has implications that can range from the small farm to Big Pharma. These developments can lessen diversity in crops and further concentrate control over agriculture in the hands of a few companies. This can create health and environmental risks that are not adequately monitored by regulatory agencies. The corporatization of life-forms, from Roundup Ready seeds to the DuPont OncoMouse, may seem at first glance no different than man's dominion over plants and animals in the past. But it is moving us toward an era where humans themselves will become not only producers but products.

Chapter Six

Taking Seriously the Claim That Genetic Engineering Could End Hunger: A Critical Analysis

Peter Rosset

In this chapter I take very seriously the oft-repeated claim that genetic engineering of crop seeds could be an important way to attack hunger, submitting it to a rigorous critical analysis. Industry and mainstream research and policy institutions often suggest that genetically engineered varieties can raise the productivity of poor Third World farmers, feed the hungry, and reduce poverty.[1] In order to address these propositions critically, we must examine the assumptions and claims that lie behind them. In order to do so, I first briefly review the notion that hunger is due to a scarcity of food, and thus that it could be remedied by producing more. I then look into the situation faced by poor farmers, including the issue of their productivity. I close by examining some of the claims made by proponents of genetic engineering, and the special risks that it may pose for poor farmers.

Food Availability and Hunger

Global data shows that there is no relationship between the prevalence of hunger in a given country and its population. For every densely populated and hungry nation such as Bangladesh or Haiti, there is a sparsely populated and hungry nation such as Brazil or Indonesia. In fact, per capita food production increases over the past four decades have far outstripped human population growth. The world today produces more food per inhabitant than ever before. Enough is available to provide 4.3 pounds for every person every day, including 2.5 pounds of grain, beans, and nuts; about a pound of meat, milk, and eggs; and another of fruits and vegetables. This is more than enough for a healthy, active life. The real causes of hunger are poverty, inequality, and lack of access. Too many people are too poor to buy the food that is available (but often poorly distributed) or lack the land and resources to grow it themselves.[2] At this level of macroanalysis, then, it should be clear that we most definitely do not need more food in order to end hunger. Thus, at a global scale, improved crop production technology of any kind is unlikely to help.

However, this may not be true in all cases of individual countries, or regions within countries, where per capita food production figures and food availability may lag behind global averages. Thus we must take seriously the notion that in some cases (e.g., parts of sub-Saharan Africa) we may have to address the productivity of poor farmers who grow foodstuffs for consumption in regional and national markets, in order to effectively combat hunger.

When we speak of these national markets, we find that small and peasant farmers, despite their disadvantaged position in society, are the primary producers of staple foods, accounting for very high percentages of national production in most Third World countries. This sector, which is so important for food production, is itself characterized by poverty and hunger, and in some cases, lagging agricultural productivity. If these problems are to be addressed by a proposed solution—genetic engineering in this case—we must begin with a clear understanding of their causes. If the causes lie in inadequate technology, then a technological solution is at least a theoretical possibility. Thus let me begin by examining the conditions faced by peasant producers of staple foods in most of the Third World.

Historical Background

The history of the Third World since the beginning of colonialism has been a history of unsustainable development. Colonial landgrabs pushed rural food-producing societies off the lands most suitable for farming, the relatively flat alluvial or volcanic soils with ample, but not excessive, rainfall (or water for irrigation). These lands were converted to production for export in the new global economy dominated by the colonial powers. Instead of producing staple foods for local populations, they became extensive cattle ranches or plantations of indigo, cacao, copra, rubber, sugar, cotton, and other highly valued products. Where traditional food producers had utilized agricultural and pastoral practices developed over thousands of years to be in tune with local soil and environmental conditions, colonial plantations took a decidedly short-term view toward extracting the maximum benefit at minimal costs, often using slave labor and production practices that neglected the long-term sustainability of production.[3]

Meanwhile, local food producers were either enslaved as plantation labor or displaced into habitats that are marginal for production. Precolonial societies had used arid areas and desert margins only for low-intensity nomadic pastoralism; had used steep slopes only for low population density, long-fallow-shifting-cultivation (or sophisticated terracing in some cases); and had used rain forests primarily for hunting and gathering (with some agroforestry)—all practices that are ecologically sustainable over the long term. But colonialism drove farming peoples—accustomed to the continuous production of annual crops on fertile, well-drained soils with good access to water—en masse into these marginal areas. Whereas precolonial cultures had never considered these regions to be suitable for high population densities or intensive annual cropping, in many cases they were henceforth to be subject to both. As a result, forests were felled and many fragile habitats were subject to unsustainable production practices, in this case by poor, newly destitute and displaced farmers, just as the favored lands were being degraded by continuous export cropping at the hands of Europeans.[4]

National liberation from colonialism did little to alleviate the environmental and social problems generated by this dynamic; in fact, the situation worsened in much of the Third World. Postcolonial national elites came to power with strong linkages to the global export-oriented economy, often, indeed, connected to former colonial powers. The period of national liberation, extending over more than a century, corresponded with the rise of capitalist market and production relations on a global scale and, in particular, with their penetration of Third World economies and rural areas. New exports came to the fore, including coffee, bananas, ground nuts, soybeans, oil palm, and others, together with new, more capitalistic (as opposed to feudal or mercantile) agroexport elites. This was the era of modernization, whose dominant ideology was that bigger is better. In rural areas that meant the consolidation of farmland into large holdings that could be mechanized, and the notion that the "backward and inefficient" peasantry should abandon farming and migrate to the cities where they would provide the labor force for industrialization. This ushered in a new era of land concentration in the hands of the wealthy, and drove the growing problem of landlessness in rural areas. The landless rapidly became the poorest of the poor, subsisting as part-time seasonal agricultural or day laborers, or as sharecroppers, or migrating to the agricultural frontier to fell forests for homesteads. Also among the poor were the "land poor": renters of small plots, squatters, or legal owners of parcels too small or too infertile to adequately support their families.[5]

Thus, rural areas in the Third World are today characterized by extreme inequalities in access to land, in security of land tenure, and in the quality of the land farmed. These inequalities underlie equally extreme inequities in wealth, income, and living standards. The poor majority are marginalized from national economic life, as their meager incomes make their purchasing power insignificant.[6]

This creates a vicious circle. The marginalization of the majority leads to narrow and shallow domestic markets, so landowning elites orient their production to export markets where consumers do have purchasing power. By doing so, elites have ever less interest in the well-being or purchasing power of the poor at home, as the poor are not a market for them, but rather a cost in terms of wages to be kept as low as possible. By keeping wages and living standards low, elites guarantee that healthy domestic markets will never emerge, reinforcing export orientation. The result is a downward spiral into deeper poverty and marginalization, even as national exports become more "competitive" in the global economy. One irony of our world, then, is that food and other farm products flow *from* areas of hunger and need *to* areas where money is concentrated, in northern countries.[7]

The same dynamic drives environmental degradation. On the one hand, rural populations have historically been relocated from areas suitable for farming to those less suitable, leading to deforestation, desertification, and soil erosion in fragile habitats. This process continues today, as the newly landless continuously migrate to the agricultural frontier.

On the other hand, the situation is no better in the more favorable lands. Here the better soils of most nations have been concentrated into large holdings used for mechanized, pesticide- and chemical fertilizer–intensive, monocultural production for export. Many of our planet's best soils—which had earlier been managed sustainably for millennia by precolonial traditional agriculturalists—are today being rapidly degraded, and in some cases abandoned completely, in the short-term pursuit of export profits and competitiveness. The productive capacity of these soils is dropping rapidly due to soil compaction, erosion, waterlogging, and fertility loss, together with growing resistance of pests to pesticides and the loss of in-soil and aboveground functional biodiversity. The growing problem of "yield decline" in these areas has recently been recognized as a looming threat to global food production by a number of international agencies.[8]

Structural Adjustment and Other Macropolicies

As if that were not enough, the past three decades of world history have seen a series of changes in national and global governance mechanisms that have, in their sum, eroded the ability of governments in southern nations to manage national development trajectories with a view to the broad-based human security of their citizens. In addition, these changes have critically weakened the ability of govenments to ensure the social welfare of poor and vulnerable people, to achieve social justice, to guarantee human rights, and to protect and manage their natural resources sustainably. These changes in governance mechanisms have been made within a paradigm that sees international trade as the key resource for promoting economic growth in national economies, and sees that growth as the solution to all ills.[9]

In order to make way for increased import/export activity and export-promoting foreign investment, structural adjustment programs (SAPs), regional and bilateral trade agreements, as well as GATT (General Agreement on Tariffs and Trade) and World Trade Organization (WTO) negotiations have all shifted the balance of governance over national economies away from governments and toward market mechanisms and global regulatory bodies like the WTO. Southern governments have progressively lost the majority of the management tools in their macroeconomic policy toolboxes. They have been forced to drastically cut government investment through deficit-slashing requirements, to unify exchange rates, devalue and then float currencies, to virtually eliminate tariff and nontariff import barriers, to privatize state banks and other enterprises, and to slash or eliminate subsidies of all kinds, including social services and price supports for small farmers. In most cases, either in preparation for entering trade agreements, or with international financial institution (IFI) funding and/or guidance, governance over land tenure arrangements has followed suit, with privatization, land markets and market mechanisms coming to the fore, in search of greater investment in agricultural sectors.[10]

While such changes have in some cases created new opportunities for poor people to exploit new niche markets in the global economy (organic coffee, for example), they have for the most part undercut both government-provided social safety nets and guarantees, and traditional community management of resources and cooperation in the face of crises. The majority of the poor still live in rural areas, and these changes have driven many of them to new depths of crisis in their ability to sustain their livelihoods. Increasingly, they have been plunged into an environment dominated by global economic forces, where the terms of participation have been set to meet the interests of the most powerful. Small farmers find the prices of the staple foods they produce dropping below the cost of production, in the face of cheap imports freed from tariffs and quotas. They are increasingly without the subsidized credit, marketing, and prices that once helped support their production, and with communal land tenure arrangements under attack from legal reforms and private sector investors. The result is the declining productivity of small farmers who produce food for domestic consumption, especially in regions such as sub-Saharan Africa.[11]

Lagging Productivity

Third World food producers demonstrate lagging productivity not because they lack "miracle" seeds that contain their own insecticide or tolerate massive doses of herbicide, but because they have been displaced onto marginal, rain-fed lands, and face structures and macroeconomic policies that are increasingly inimical to food production by small farmers. When development banks are privatized by SAPs, credit is withdrawn from small farmers. When SAPs cancel subsidies for inputs, small farmers stop using them. When price supports end, and domestic markets are opened to surplus food dumped by northern countries, prices drop and local food production becomes unprofitable. When state marketing agencies for staple foods are replaced by private traders, who prefer cheap imports or buying from large wealthy farmers, small farmers find there are no longer any buyers for what they produce. These, then, are the true causes of low productivity. In fact, in many parts of the Third World, especially in Africa, *farmers today produce far less then they could with currently available know-how and technology,* because there is no incentive for them to do so—there are only low prices and few buyers. No new seed, good or bad, can change that, and thus it is extremely unlikely that, in the absence of urgently needed structural changes in access to land and in agricultural and trade policies, genetic engineering could make any dent in food production by the world's poorer farmers.[12]

When seen in this light, it should be clear that genetic engineering is tangential at best to the conditions and needs of the farmers we are told it will help and in no way addresses the principal constraints they face. But tangential is a far cry from "bad." Now I turn to the question of whether genetically engineered crops are simply irrelevant to the poor, or if they might actually pose a threat to them. First we must ask about the actual circumstances of peasant farming.

A Complex, Diverse, and Risk-Prone Agriculture

Because peasant farmers have historically been displaced, as described earlier, into marginal zones characterized by broken terrain, slopes, irregular rainfall, little irrigation, and/or low soil fertility; and because they are poor and are victimized by pervasive anti-poor and anti-small farmer biases in national and global economic policies, their agriculture is best characterized as complex, diverse, and risk-prone.[13]

In order to survive under such circumstances, and to improve their standard of living, they must be able to tailor agricultural technologies to their variable but unique circumstances, in terms of local climate, topography, soils, biodiversity, cropping systems, market insertion, resources, and the like. For this reason, such farmers have over millennia evolved complex farming and livelihood systems that balance risks —of drought, of market failure, of pests, and so on—with factors such as labor needs versus availability, investment needed, nutritional needs, and seasonal variability. Typically, their cropping systems involve multiple annual and perennial crops, animals, fodder, even fish, and a variety of foraged wild products.[14]

Repeating the Errors of Top-Down Research

Such farmers have rarely benefited from "top-down" formal institution research and Green Revolution technologies.[15] Any new strategy to truly address productivity and poverty concerns will have to meet the need for multiple suitable varieties. Peasant farmers typically plant several different varieties on their land, tailoring their choice to the characteristics of each patch, whether it has good drainage or bad, is more fertile or less fertile then the rest, and so on. However, such varieties cannot be easily developed with current research and extension structures and methods—the same structures that biotech proponents use for genetically engineered varieties. Formal research methods are not able to handle the vast complexity of physical and socioeconomic conditions in most Third World agriculture. This stems from the discrepancy between hierarchical research and extension systems, which value monocultural "yield" above all else, and complex rural realities. The result of the mismatch is that numerous variables important to farmers have to be reduced in order to produce new technologies. Measured in a few variables, new seeds are perceived by researchers to be better than old ones, and these same researchers are puzzled when farmers fail to adopt the new seeds widely.[16]

In reality, seeds have multiple characteristics that cannot be captured by a single yield measure, as important as this measure may be, and farmers have multiple site-specific requirements for their seeds, not just controlled-condition high yields. These interconnections stand in direct contrast to formal breeding procedures where varieties are selected individually for discrete traits, then crossed to combine these individual traits. According to Jiggins and co-authors, high-yielding variety trials in sub-Saharan Africa (SSA) show "larger variations, for both 'traditional' and 'improved,' *among* farmers and *between* years, than the mean differences between 'traditional' and 'improved' yields in a single year. There is indeed overwhelming evidence throughout SSA that the yield response to fertilizer and improved varieties, soil management and other practices is highly site-, soil-, season-, and farmer-specific."[17]

Given such conditions, the inescapable conclusion is that a different approach—participatory breeding by organized farmers themselves, which takes into account the multiple characteristics of both seed varieties and farmers—is essential; miracle seeds will not just be developed in laboratories and on research stations and then effortlessly distributed to farmers.[18] Genetic engineering is the very antithesis of participatory, farmer-led research. Proponents of genetically engineered varieties are repeating the very top-down errors that led first-generation Green Revolution crop varieties to have low adoption rates among poorer farmers.

Yet many would argue that the possibility of delivering enhanced nutrition to the poor should outweigh such concerns—for example, in the case of the famous Golden rice, which has been engineered to contain additional beta carotene, the precursor of vitamin A.

Enhanced Nutrition?

The suggestion that genetically altered rice is the proper way to address the condition of 2 million children at risk of vitamin A deficiency–induced blindness reveals a tremendous naïveté about the reality and causes of vitamin and micronutrient malnutrition. If one reflects upon patterns of development and nutrition, one must quickly realize that vitamin A deficiency is not best characterized as a problem, but rather as a *symptom*, a warning sign. It warns us of broader dietary inadequacies associated both with poverty and with agricultural change from diverse cropping systems toward rice monoculture. People do not present with vitamin A deficiency because rice contains too little vitamin A, or beta carotene, but rather because their diet has been reduced to rice and almost nothing else, and they suffer many other dietary illnesses that cannot be addressed by beta carotene, but that *could* be addressed, together with vitamin A deficiency, by a more varied diet. A magic-bullet solution that places beta carotene into rice—with potential health and ecological hazards—while leaving poverty, poor diets, and extensive monoculture intact is unlikely to make any durable contribution to well-being. To use the words of Dr. Vandana Shiva, such an approach reveals *blindness* to readily available solutions to vitamin A deficiency–induced blindness, including many ubiquitous leafy plants that when introduced (or reintroduced) into the diet provide both needed beta carotene *and* other missing vitamins and micronutrients.[19]

Yet it is clear that the biotech juggernaut is moving ahead at full speed. What then are the risks associated with "forcing" transgenic (genetically engineered) varieties into complex, diverse, and risk-prone circumstances?

Risks for Poor Farmers

When transgenic varieties are used in such cropping systems, the risks are much greater than in Green Revolution, large, wealthy farmer systems, or farming systems in northern countries. The widespread crop failures reported for transgenics (e.g., stem splitting, boll drop, etc.)

pose economic risks that can affect poor farmers much more severely than wealthy farmers. If consumers reject their products, the economic risks are higher the poorer one is. Also, the high costs of transgenics introduce an additional anti-poor bias into the system.[20]

The most common transgenic varieties available today are those that tolerate proprietary brands of herbicides, and those than contain insecticide genes. Herbicide-tolerant crops make little sense to peasant farmers who plant diverse mixtures of crop and fodder species, as such chemicals would destroy key components of their cropping systems.[21]

Transgenic plants that produce their own insecticides—usually using the Bt gene—closely follow the pesticide paradigm, which is itself rapidly failing due to pest resistance to insecticides. Instead of the failed "one pest–one chemical" model, genetic engineering emphasizes a "one pest–one gene" approach, shown over and over again in laboratory trials to fail, as pest species rapidly adapt and develop resistance to the insecticide present in the plant. Bt crops violate the basic and widely accepted principle of integrated pest management (IPM)—that reliance on any single pest management technology tends to trigger shifts in pest species or the evolution of resistance through one or more mechanisms. In general, the greater the selection pressure across time and space, the quicker and more profound the pests' evolutionary response. Thus IPM approaches employ multiple pest control mechanisms, and use pesticides minimally, only in cases of last resort. An obvious reason for adopting this principle is that it reduces pest exposure to pesticides, retarding the evolution of resistance. But when the product is engineered into the plant itself, pest exposure leaps from minimal and occasional to massive and continuous, dramatically accelerating resistance. Most entomologists agree that Bt will rapidly become useless, both as a feature of the new seeds and as an old standby natural insecticide sprayed when needed by farmers who want out of the pesticide treadmill. In the United States, the Environmental Protection Agency has mandated that farmers set aside a certain proportion of their land as a "refuge," where non-Bt varieties are to be planted, in order to slow down the rate of evolution by insects of resistance. Yet it is increasingly unlikely that poor, small farmers in the Third World will plant such refuges, meaning that resistance to Bt could occur much more rapidly under such circumstances.[22]

At the same time, the use of Bt crops affects nontarget organisms and ecological processes. Recent evidence shows that the Bt toxin can affect beneficial insect predators that feed on insect pests present on Bt crops, and that windblown pollen from Bt crops found on natural vegetation surrounding transgenic fields can kill nontarget insects. Small farmers rely for insect pest control on the rich complex of predators and parasites associated with their mixed cropping systems. But the effect on natural enemies raises serious concerns about the potential of the disruption of natural pest control, as polyphagous predators that move within and between mixed crop cultivars will encounter Bt-containing nontarget prey throughout the crop season. Disrupted biocontrol mechanisms may result in increased crop losses due to pests or to the increased use of pesticides by farmers, with consequent health and environmental hazards.[23]

The fact that Bt retains its insecticidal properties after crop residues have been plowed into the soil, and is protected against microbial degradation by being bound to soil particles, persisting in various soils for at least 234 days, is of serious concern for poor farmers who cannot purchase expensive chemical fertilizers, and who instead rely on local residues, organic matter, and soil microorganisms (key invertebrate, fungal, or bacterial species) for soil fertility, which can be negatively affected by the soil-bound toxin.[24]

When the Bt genes fail, what would poor farmers be left with? It is entirely possible that they would face the serious rebound of pest populations freed of natural control by the impact Bt had on predators and parasites, and reduced soil fertility because of the impacts of Bt crop residues plowed into the ground.[25] These are farmers who are already risk-prone, and Bt crops would likely increase that risk.

In the Third World there will typically be more sexually compatible wild relatives of crops present, making pollen transfer to weed populations of insecticidal properties, virus resistance, and other genetically engineered traits more likely, with possible food chain and superweed consequences. With massive releases of transgenic crops, these impacts are expected to scale up in those developing countries that constitute centers of genetic diversity. In such biodiverse agricultural environments, the transfer of coding traits from transgenic crops to wild or weedy populations of these taxa and their close relatives is expected to be higher. Genetic exchange between crops and their wild relatives is common in traditional agroecosystems and transgenic crops are bound to frequently encounter sexually compatible plant relatives, therefore the potential for "genetic pollution" in such settings is inevitable.[26]

There is potential for vector recombination to generate new virulent strains of viruses, especially in transgenic plants engineered for viral resistance with viral genes. In plants containing coat protein genes, there is a possibility that such genes will be taken up by unrelated viruses infecting the plant. In such situations, the foreign gene changes the coat structure of the viruses and may confer properties such as changed method of transmission between plants. The second potential risk is that recombination between RNA virus and a viral RNA inside the transgenic crop could produce a new pathogen leading to more severe disease problems. Some researchers have shown that recombination occurs in transgenic plants and that under certain conditions it produces a new viral strain with altered host range.[27] Crop losses caused by new viral pathogens could have a more significant impact on the livelihoods of poor farmers than they would for wealthier farmers who have ample resources to survive poor harvests.

In sum, these and other risks seem to outweigh the potential benefits for peasant farmers, especially when we consider the factors that currently limit their ability to improve their livelihoods, and the proven agroecological, participatory, and empowering alternatives available to them.[28]

The Parable of the Golden Snail

It is not a lack of technology that holds such farmers back, but rather pervasive injustices and inequities in access to resources, including land, credit, market access, and other anti-poor policy biases. Two approaches make the most sense under such conditions: technologies that have pro-poor diseconomies of scale, such as agroecology,[29] and organization into social movements capable of exerting sufficient political pressure to reverse policy biases. There is little useful role that genetic engineering can play.

When a group of Filipino farmers were asked recently for their thoughts on genetically engineered rice seeds, a peasant leader responded with what might be called the Parable of the Golden Snail. It seems that rice farmers have long supplemented the protein in their diet with local snails that live in rice paddies. At the time of the Marcos dictatorship, Imelda Marcos had the idea of introducing a snail from South America that was said to be more productive and, as such, a means to help end hunger and protein malnutrition. But no one liked the taste, and the project was abandoned. The snails, however, escaped, driving the local snail species to the brink of extinction—thus eliminating a key protein source—and forcing peasants to apply toxic pesticides to keep the South American species from eating the young rice plants. "So when you ask what we think of the new genetically engineered rice seeds, we say that's easy," the leader said. "They are another Golden Snail."[30]

Next time we hear of the latest magic bullet, altruistically developed in private-sector labs for the benefit of the poor, we would do well to heed this parable, as well as to keep in mind the true causes of hunger, poverty, and lagging agricultural productivity in the Third World.

Chapter Seven

The European Response to GM Foods: Rethinking Food Governance

David Barling

This chapter reviews the European response to the large-scale introduction of genetically modified (GM) crops and their food derivatives into the European Union (EU). The inadequacies of the regulatory frameworks for GM crops and foods and for consumers in the EU are reviewed and the current efforts to revise these frameworks explained. The implications of these events for the governance of the food system are considered and some important lessons for the future put forward.

The introduction of GM crops and their food derivatives into the European food chain in the late 1990s was met with widespread public concern and opposition. While the public voiced their complaints as consumers, large corporate retailers responded by seeking non-GM sources for many of their own brand foods. The response generated a renewed debate over the inadequacies of the existing regulatory framework in the EU for GM foods and for consumer choice through the labeling provisions. The limitations of the risk assessment framework for the impacts of GM crops on the agri-environment and its biodiversity also became headline news. Within a few years of the introduction of GM crops and foods, their regulation moved from being a relatively closed and specialized policy network focusing on quite technical issues, to one in which the agenda of high politics generated much newsprint and spilled over into international trading relations.

The response of the European public was of no great surprise to students of earlier public opinion surveys of the public's attitudes to GM foods, and of more qualitative analyses of the underlying values the public might bring to assessing the suitability of GM foods. An analysis of some of this earlier work illustrates that the warning signs were there for those who chose to look, both industry and government regulators alike.

The Large-Scale Entry of GM Crops into the EU: The Market Responds

The first large-scale planting of GM crops began in North America in 1996. By 1998, farmers in the large crop-growing areas of countries such as the United States, Canada, and Argentina grew commercially an estimated 27.8 million hectares (68.7 million acres) of GM crop varieties, increasing to 39.9 million hectares in 1999.[1] This first generation of GM crops was cleared by government authorities for field characteristics determined by bacterial genes, such as tolerance to the broad-spectrum herbicides glyphosate and glufosinate ammonium, and insect resistance due to the insertion of genes from the soil bacterium *Bacillus thuringiensis* (Bt). These GM crops were novel extensions of the agrochemical technology inputs that promised potential agronomic benefits to farmers, such as reducing the number of herbicide sprays during crop cultivation and more effective immediate pest control. The large agricultural biotechnology companies were estimated to have spent approximately $2000 million on research and development by 1995, while sales of agricultural biotechnology products amounted to just $100 million in that year.[2] Hence, there was a pressing need to realize these investments with increased sales as the first large-scale plantings of GM crops began.

In 1996, Monsanto's Roundup Ready soybean, containing bacterial genes for glyphosate tolerance and a marker gene resistant to the antibiotic ampicillin, received an EU marketing licenses for grain importation, storage, and use in agriculture. This was initially seen as a turning point for the industry, by critics and supporters alike, as it allowed for the large-scale importation of GM soybeans and derivatives into the European food market.[3] It was estimated that something like 60 percent of foodstuffs on sale in Europe would contain GM soybean derivatives. In 1997, Novartis Maximizer GM Bt corn, which included novel genes for resistance to the European corn borer, marker genes for tolerance to glufosinate and resistance to the antibiotic ampicillin, and the cauliflower mosaic gene promoter, was licensed for importation and agricultural use. Subsequently, in 1998 some 15,000 hectares of Bt corn were grown in Spain and 1,000 hectares in France.[4]

The first large-scale shipment from the United States of GM soybean mixed in with nonmodified grain reached Antwerp in 1996, where Greenpeace organized an unsuccessful attempt to prevent the unloading of the cargo. The protests did succeed in raising public awareness, notably amongst northern continental consumers, of the large-scale entry of GM commodities into the European food chain. The disquiet amongst the German public led some of the main food processors and distributors, including Unilever and Nestlé, to seek to remove soy oil derived from these modified soybeans from their ingredients at the end of 1996.[5] However, they refused at this stage to remove the oil from their products sold in other European countries.

The publicity attached to the imports of the soybeans and corn was reinforced by a renewed focus on the potential impacts on biodiversity and farming practice in the European agri-environment from herbicide- and pesticide-modified GM crops. Monsanto had launched a high-profile public relations campaign in the United Kingdom (U.K.) in the summer of 1998 to persuade the public of the benefits of food biotechnology. For example, in a full-page advertisement placed in the press, Monsanto extolled the belief that "biotechnology is one way to cut down on the amount of pesticides used in agriculture."[6] However, nongovernmental organizations (NGOs) who were keen to expand the debate beyond the narrow confines of environmental policy networks and to involve the wider public swiftly countered such assertions. As well as providing this role of counterexpertise, some NGOs used symbolic protest events to amplify the message, hopeful that latent public concerns would turn to more vocal support, such as the destruction of fields of GM crop trials. The public in turn voiced their concerns as food consumers. The sluggish response of the regulatory process was quickly outpaced by the further response of the key commercial players in the food chain as they picked up the cues from their customers.

By the end of 1998, the public discontent at the entry of these GM products into the food chain consequently spread to other major European nations, notably France and the U.K. The public's unease in the U.K. reached fever pitch in the early months of 1999, as was reflected by campaigns on the issue of GM foods by the *Daily Express* and the *Daily Mail*, both middle-range tabloid newspapers. The sensitivity of the U.K. market was reflected in the case of a GM tomato paste that had been launched in both the Sainsbury and the Safeway supermarket chains in the U.K. in 1996. The launch of this product was seen as an example of good practice for a new technology product entry, with full labeling and explanatory leaflets. Initially, it outsold its nonmodified alternative, which it undercut in price. However, by the end of the decade, the supermarkets had withdrawn the product due to falling sales. Sainsbury responded to the increased media reports about GM food in early 1999 by opening a dedicated customer call line on the subject. They received three hundred calls in the first four hours, and reacted accordingly.[7]

Initial pleas from European food retailers and processors for the segregation of modified from nonmodified soybeans for the European market were met with resistance by the large U.S. grain companies and the American Soybean Association.[8] The U.K. frozen food retailer, Iceland, responded by leading a search for nonmodified sources of soybeans for its own brand foods. It sourced nonmodified soybean derivatives from Canada and Brazil. The company sought to verify their supplies using the detection methodology for GM DNA of polymerase chain reaction (PCR) technology, developed by a U.S. company, which they then brought over to labs in the U.K. However, the reliability of these detection methods was not complete. In response, Iceland supported these methods with a clear audit trail of testing through the different stages of the food chain, from the field through to the different processing and manufacturing phases. In March 1998, the company was able to announce its own brand foods as being non-GM. To quote Bill Wadsworth, technical director of Iceland, who helped to pioneer the search for a non-GM soybean supply for Europe: "The only way that any European consumer was given a choice was because we fundamentally broke the supply chain. We set up a totally unique supply chain."[9]

The major supermarket retailers in the U.K. and northern Europe soon followed Iceland's lead. In March 1999, a consortium of Sainsbury, Marks & Spencer, Carrefour (France), SuperQuinn (Ireland), Effelunga (Italy), Migros (Switzerland), and Delhaize (Belgium) was announced. Their intention was to create a market presence with enough buying power to ensure the maintenance of a non-GM supply chain, fully tested and audited for their own brand products.[10] The major processors also followed suit, under pressure from the retailers, as did some of the major restaurant and catering groups. A survey by

Friends of the Earth found that in March 2000 most of the world's top twenty-six food manufacturers who sold in the European market had adopted non-GM policies for that market.[11] The managing director of Cargill in the U.K., speaking in late 1999, considered that for the next five years the food industry would "remain where it is . . . i.e., that it will not receive GM ingredients."[12]

Nonetheless, the emergence of a supply chain for non-GM commodities such as soybeans and corn remains a challenge, as the widespread planting of GM varieties in North America is maintained, and other non-GM sources such as Brazil seem likely to allow commercial planting of GM soybeans. The problems of non-GM contamination by GM equivalents exist from the production of seed varieties through to the farm stages of planting, harvesting, and storage, and movement off farm to the grain merchants. The storage, handling, and transportation of the grain, from elevator to silo to port to container and into the processing chain overseas, provide further challenges to effective segregation. The contamination of Advanta's Hyola oilseed rape by a GM variety and sold through Europe (traced back to crop production in Canada) and of human food by Aventis's GM StarLink corn (approved only for animal feed use in the United States) in 2000 illustrates the ongoing difficulties of segregation.

At the production end of the food system, the first generation of GM products reflected the growing corporate concentration that was developing in the so-called life sciences sectors. The life sciences strategy was to bring together pharmaceuticals, biotechnology, agribusiness, food, chemicals, cosmetics, and energy (e.g., ethanol) all under one corporate umbrella. The large agrochemical companies were rapidly merging and integrating in the 1990s both horizontally and vertically. For example, Aventis was formed from the merger of Hoescht (which included AgrEvo) and Rhone Poulenc. Novartis emerged from the merger of Ciba-Geigy and Sandoz, and Astra Zeneca was a merger of a pharmaceutical with a crop science company. In the United States, Monsanto and DuPont, as well as Novartis and Astra Zeneca from Europe, were rapidly absorbing seed companies to exploit their germ plasm holdings.[13] In addition, the agricultural biotechnology corporations had entered into strategic alliances with the large agribusinesses that in turn controlled large sections of the grain commodity gathering and distribution markets in the United States and for overseas export. Monsanto formed an alliance with Cargill, and Novartis with Archer Daniels Midland (ADM).[14] With the widespread rejection of these GM products in the EU and other important export markets such as Japan, the value of agricultural biotechnology within the large "life sciences" corporations came under scrutiny. Further mergers have occurred (e.g., Monsanto with Pharmacia & Upjohn), and the possible float off of the agricultural biotechnology arms from some of these large corporations has been further discussed.

In effect, in Europe a market-led private system of regulation with regard to GM food has emerged. This system has been enabled both by the corporate concentration in different stages of the food chain (at retailing, processing, and, to a lesser extent, catering), and to some extent by the improvements in detection technology. At the production and commodity distribution end of the food system, corporate concentration has facilitated the development of agricultural biotechnology. Ironically, corporate concentration at the retail and food-processing ends of the food system facilitated the European consumers' rejection of the technology. The large supermarkets extended their position as brokers in the food system based on their gatekeeper position to mass consumer markets in the EU, and reinforced by tight systems of standards and hygiene along the supply chain.[15]

The rejection of GM foods in the EU at the end of the 1990s was not a total surprise to many observers of the food system. There were clear signs among studies of public attitudes and values that were inimical to the spread of biotechnology into food and agriculture. The campaigns of the NGOs and the press found a resonance among the sceptical and uneasy public, wary of an increasingly risk-prone industrial agrifood system. GM foods became part of a litany of concerns regarding the food system, adding to pesticide residues, industrial food additives, salmonella and *E. coli* contamination, and BSE (bovine spongiform encephalopathy, or mad cow disease).

The European Public and GM Food: The Early Warning Signs

The reaction of the European public was anticipated by earlier quantitative surveys and more qualitative studies of public attitudes and values concerning biotechnology and its applications, such as GM foods and crops that had been conducted within Europe. The warning signs had been there, but had been largely ignored, or misread, by industry and governments alike.

One such sign was contained in the *Eurobarometer*, EU-wide surveys of public attitudes carried out on behalf of the European Commission.

Random surveys on European citizens' attitudes to biotechnology had been carried out in 1991, 1993, and 1997.[16] Findings of the 1997 survey, "The Europeans and Modern Biotechnology," amplified concerns raised in earlier quantitative studies. There was a correlation among respondents between a greater objective knowledge and both optimistic and pessimistic views of the anticipated effects of biotechnology. Hence, greater knowledge did not necessarily bring either increased support for biotechnology or greater opposition. Perceptions of risk were higher among those with greater objective knowledge and those who had discussed biotechnology over recent months, but such perception was low among those with little knowledge. Overall, the findings suggested a lack of confidence in self-regulation by industry, but also a relative lack of trust and confidence in the effectiveness of EU and national regulation and institutions, suggesting worrying implications for the legitimacy of the governance of modern biotechnology.

In the 1997 survey, the question to identify the most reliable source of information on biotechnology elicited the following responses:

- Consumer organizations (58%)
- Environmental protection organizations (56%)
- Schools/universities (35%)
- Public authorities (18%)
- Industry (7%)

Finally, the majority of the respondents (74%) favored labeling of GM foods.[17]

A problem for such surveys of attitudes to biotechnology has been the lack of prior knowledge and understanding of the technology among the public. However, respondents can, and do, draw on their existing cognitive frameworks to formulate their attitudes on the technology and its applications.[18] Hence, in addition to these quantitative surveys, more qualitative attempts have been made to assess citizens' attitudes to GM foods by seeking to relate consumer attitudes to their more deeply held concerns and values. Such surveys were conducted in the Netherlands[19] and the U.K.[20] The key values that emerged across these studies were based around perceptions of trust, choice, need, and care for a sustainable society, which included conceptions of natural balance.

Lack of trust in the regulatory process and the credibility of regulations emerged from research in the U.K. and was linked to a weak sense of agency in the regulatory process.[21] Conversely, in the Netherlands, where the respondents did feel trust in the regulatory process and authorities, they also felt that they had control. This trust, however, could be undermined by fear, which could be induced by lack of knowledge of food production and how to interpret information on the packaging.[22]

The notion of choice was a complex issue. Freedom of choice was linked to variation in products and variation in prices, as well as consumer knowledge about the interpretation of information on packages of foodstuffs and the production methods. Because of a lack of knowledge and uncertainty over the effects on health and natural balance, biotechnology undermines the public's perception of their freedom of choice somewhat and engenders some fear and insecurity.[23] On consumer choice, the low price of a food was a key determinant, but was also seen as an economic constraint against real personal choice. Labeling was seen to be of limited utility because of the desire to have information about the social context of the decision to produce a GM food. Choice as a citizen was also an important factor, but here, perhaps in contrast to citizens in the United States, there was resignation and a lack of a sense of personal agency.[24] Consumers are also at pains to discriminate between classes of product and the different issues, negative and positive, that these raise.[25] For example, whether the use of herbicide was increased or reduced was seen as an important criterion for assessing GM crops.[26]

The notion of a sustainable society was embraced by converging opinions concerning how modern biotechnology impacts the natural balance, trust, health, social dissipation, and Third World problems, and its intrinsic usefulness or necessity. "Natural balance" covered recognition of potentially positive implications of biotechnology, including possibly cleaner, less chemical dependent, and more efficient food production; protecting the environment; using fewer raw materials; and the preservation of rare species through cloning. Negative effects were also identified: decline in crop diversity as supercrops predominate food production; overproduction and its attendant threats to the environment and ecosystems; and irreversibility of harm, for example, upon different trophic levels in the natural food chain.[27]

The qualitative studies pointed to a set of values that the public would be likely to apply to the entry of the products of biotechnology into the food chain. Such values were mobilized with the entry of GM foods and derivatives into the European food chain, as NGOs seized on these latent concerns as a platform for opposition to GM foods and for revision of the regulatory frameworks governing them. While the super-markets responded swiftly to the market signals, the regulators struggled to keep up, underlining the EU public's lack of confidence and trust. Nonetheless, the regulators have sought to address the trust deficit that had emerged, not least in an effort to seek to restore their own legitimacy in the eyes of the European public.

Regulatory Responses in the EU

By 1998, the EU regulatory framework for GM crops and foods (but not for GM animal feed and GM seed production) was already seen as too limited. The environmental impacts of the deliberate release of genetically modified organisms (GMOs) such as GM crops were regulated under a directive passed in 1990 (directive 90/220/EC). However, this directive was under pressure to be revised within a few years of its inception, and the European Commission announced its revision in 1994.

At the end of the 1990s, two countervailing reform pressures were at work. Industry desired a more streamlined and quicker approval process for the marketing of GM crops. The large agricultural biotechnology companies had found a receptive audience within parts of the European Commission, who saw the development of biotechnology as a driver of economic growth and industrial competitiveness in the world market. Such economic growth was viewed as consolidating the legitimacy of the EU in the minds of the European public. This market-driven approach accepted the need for regulation for harmonization of a common trading market for the EU, but sought regulation with a lighter touch that would not act as a disincentive to industries based on new technologies. The large agricultural biotechnology corporations interests' had formed an effective policy advocacy position alongside receptive directorate-generals (DGs) (the departments) of the Commission. The corporations lobbied as the Senior Advisory Group on Biotechnology and subsequently as EuropaBio (incorporating national associations of smaller biotech companies).

Conversely, the coordination of the implementation of the deliberate release directive was based in DG Environment, which managed the market approval process and the inputs of the competent authorities of the EU member states. Environmental pressure groups and the environment committee of the European Parliament were also part of this environmental protection–dominated policy network. The reform impulse from this network, led by certain member states and lobbied by the environmental groups, was to strengthen the detail in the directive, notably in terms of the risk assessment criteria for GM crops that should be met before granting market approval.[28]

Risk Assessment Directives

The risk assessment boundaries under the directive were vague and unclear, and had generally failed to take adequate account of the possible wider ecological effects of the release of GMOs, in particular with regard to potential impacts upon local and regional biodiversity in Europe. Also, the agricultural impacts of specific GM crops (e.g., modified for herbicide resistance) were often ignored in the risk assessment process under the regulatory regime's procedures. Consequently, although certain GM varieties of corn and oilseed rape gained regulatory approval under the EU process, they were subsequently denied market access in individual member states, putting further stress on the regulatory regime.

The resulting debates featured epistemic tensions between the differing perspectives offered by molecular biologists and ecologists. There was an inherent tension within a regulatory regime that attempted to treat differing crop ecosystems within European agriculture as a single environment in order to meet the needs of a single economic market. Partly as a result, no products have been authorized since 1998, with the Council of Ministers declaring an effective moratorium on new approvals in June 1999. A blocking minority of member states (Denmark, France, Greece, Italy, and Luxembourg) declared that they would continue to impose this block on any further approvals until the passage of the revised directive.[29]

The revised directive (2001/18) was adopted in February 2001 and included the requirement for much more specific principles and criteria for risk assessment to be included in the application to market a GM crop. The possible effects—whether direct or indirect, immediate or delayed—upon human health and the environment are included. The environmental impacts include those on biodiversity and farm management practices. The use of antibiotic-resistant genes that may be used in medical and veterinary medicine are to be phased out from insertion into GMOs by the end of 2004. Additionally, the directive emphasized the use of the precautionary principle in the regulation.

However, the directive lacked strict legal liability provisions regarding polluters. The European Parliament also had sought an amendment to allow for socioeconomic factors to be included in the risk assessment steps applied to GMO releases. This would have been beneficial for organic farmers, in particular, and their needs are not explicitly addressed in the directive.[30] However, the Commission (in 2001) also drafted a proposal for regulating the standards of GM seed production in the EU, with the aim of reducing the risks of contamination of non-GM seeds. Following the adoption of 2001/18 in February 2001, six member states (Austria, Denmark, France, Greece, Italy, and Luxembourg) declared they would oppose any new GMO approvals until more complete and full labeling of GM products was achieved. This extended the moratorium on new approvals.[31]

In 1992, the EU Commission had introduced the Regulation on Novel Foods and Food Ingredients (258/97), but it did not become law until early 1997. The main areas contested were the scope of the regulation and the labeling provisions. Environmental risk assessment under 90/220 was retained. However, the regulation excluded food additives as well as extraction solvents and flavorings as they fell under existing product regulations. Under the principle of "substantial equivalence," foods and ingredients deemed to be equivalent to conventional foods were allowed to be approved under simplified procedures.[32]

Labeling Issues

Labeling was an important area of disagreement during the passage of the regulation. Germany, Denmark, and the Netherlands opposed the Commission's industry-supported proposals for voluntary labeling. The final regulation allowed for labeling only for live GMOs, and where the composition of the final product was deemed not equivalent to an existing food or ingredient. The labeling had to indicate the characteristics or properties modified, together with the method (or process) of modification. There was also provision for labeling on ethical grounds, and where ingredients had an allergy implication for a section of the population.

However, these provisions did not end the labeling debate. Just as the Council of Ministers was finalizing the Novel Foods Regulation, the large-scale entry of GM foods and derivatives into the European market began. The prevalence of soybean derivatives in processed foods highlighted the shortcomings of the labeling provisions in the Novel Foods Regulation. It was estimated that about 60 percent of foodstuffs on sale in Europe would contain GM soybean derivatives, but would not be labeled as such.[33] Within months of the passage of the regulation, the European Commissioner for Industry admitted the labeling provisions would need revision.[34]

A string of regulations was quickly passed to extend labeling to GM soybean and corn and derivatives in processed foods. However, the extensions of the labeling regime still focused on the composition of the final product, not upon the methods of production. A regulation concerning traceability and labeling of GMOs and traceability of food and feed products from GMOs was proposed in mid-2001, some years after their introduction by the supermarkets. This offered the possibility of regulated systems for traceability of GM foods and ingredients from seed to shelf, allowing consumers to make a process-based choice of food purchase. In addition, a new regulation on GM food and feed was proposed.[35]

Compositional Analyses

The initial GM food and feed proposal would cover all products produced *from* a GMO but not products produced *with* a GMO. The former implies that a proportion of the end product, whether the food or the feed itself or one of the ingredients in either has been derived from the original GM material. The latter is produced with the assistance of a GMO, but no material derived from the GMO is present in the end product. Hence, the regulation does not include GM enzymes used in feed and food production that no longer remain in the food product, such as cheese produced with GM chymosin, nor does it include products obtained from animals fed with genetically modified feed or treated with GM medicinal products.

The new proposal would cover food and feed additives, flavorings, and highly processed products, such as oils, which were previously exempted from the Novel Foods Regulation and subsequent labeling regulations for products derived from GM corn and soybeans (and that contain no viable GM protein or DNA). A product authorization should not be granted for a single use, when a product is likely to be used for both food and feed purposes. This closes the loophole that allowed StarLink corn, which had been approved as animal feed only, to get into the human food chain. A key change from the authorization process under the earlier novel foods regulation is that in the GM food-feed proposal, the simplified notification procedure for GM foods that are substantially equivalent to existing foods is abandoned. The Commission concedes that while "the use of this regulatory short-cut for so-called 'substantially equivalent' foods is a key step in the safety assessment process of GM foods, it is not a safety assessment in itself."[36]

A Revised Regulatory Framework

The EU has been stung into revising its regulatory framework for GM crops and foods. The revisions proposed suggest some substantial changes, with further repercussions for the international food supply chain. While the European Commission has the role of developing policy solutions to ensure the operation of the European treaties, the pressure for reform came from the public of the member states. The member states' government representatives at the EU in turn promoted regulatory revision. The tightening of the risk assessment criteria in the deliberate release directive was the result of the prompting of the member states, as was the proposal for traceability of GM food and feed, and the addition of animal feed to the revised reform proposal. A major driver of the EU as a regulatory exercise is the effective operation of a single integrated market. We might wish to consider if this is an adequate basis for a sound food policy. Or, alternatively, do we need a different conceptual approach to the governance of the food system?

Contradictions and Challenges in the Food System.

The EU's proposed revisions to their regulatory framework for GM crops and food in the EU suggests that some lessons have been learned from the disastrous attempts to foist GM foods on the European public. As consumers, the public are demanding more knowledge about the source of their food. The traceability proposals for GM food and feed reflect the European public's desire to know the provenance of their food, notwithstanding the complexities of the modern food system. EU legislation will impact along the entire international food supply chain. Despite the opposition of industry in the form of the agricultural biotechnology corporations and the large grain merchants, and now the animal feed companies, the EU is only following the lead of the market.

At the near consumption end of the food supply chain, a system of private regulation in the food retailing market is already in place. The large multiple-food retailers have imposed tight traceability through their contract specifications, often in the form of standards such as hygiene and presentational quality. This is a dominant dynamic in the contemporary food supply chain. The response of Iceland and the larger supermarket chains in Europe to the consumers' desire for non-GM food was to apply this dynamic. Only here we saw a clash of competing dynamics, in the form of different commercial interests. On the one hand, the U.S.-based life sciences corporations and their agribusiness allies; on the other, the European-based retailers and caterers, with the food (and now feed) processors and manufacturers stretched across this division. The processors and manufacturers might find themselves with feet dancing across each camp as, say, Archer Daniels Midland (ADM) engages in first-stage processing, while other first-stage processors and food manufacturers (usually engaged in second-stage processing) provide supermarkets with their own brand produce. At the same time, processors and manufacturers have their own GM interests, such as having the use of GM enzymes excluded from regulation. To this extent, some of the contradictions of the contemporary food system are being played out.

The Politicization of Ag-Biotech

GM foods have become an issue of high politics. The contradictions of the modern food system are also being played out across the diplomatic table of international trade. The United States employed the rhetoric of potential trade war over the transatlantic disagreement about GM foods. The agricultural biotechnology companies lobbied the Clinton administration and it put pressure on the U.K. government under Tony Blair to seek concessions from their European partners, although to little effect.[37]

Many of the issues around GM food regulation remain on the agenda of the United Nations' FAO/WHO Codex Alimentarius Commission (Codex). The General Agreement on Trade and Tariffs Uruguay Round agreements from 1993 had set up a rules-based regime for international trade, including food and agriculture, under the World Trade Organization (WTO). The agreement gave the WTO governance over disputes concerning food standards, including the power to impose sanctions on member states deemed to have broken the rules. The advisory body on food safety and standards to the WTO is Codex. A further agreement, on technical barriers to trade, also included a rule prohibiting the imposition of labels on traded goods that discriminated against them on the basis of means of production. This may offer a potential challenge to the evolving European labeling and traceability regime through WTO disputes procedures. However, such a challenge would do little to enhance the legitimacy of the WTO and its procedures in the eyes of the European public.

A joint FAO/WHO Codex task force on food derived from biotechnology began its work in 2000, with the aim that its recommendations feed into the Codex standards-setting process. In addition to the task force, various Codex committees are looking at defining or setting standards on issues pertinent to the governance of GM foods. These include labeling and traceability systems for GM foods, the uses of the precautionary principle, and the role of other legitimate factors in the risk analysis process. The placing of these debates in the Codex process is a controversial one. Studies of Codex have found decision making to be labyrinthine, far removed from public scrutiny. Public interest group participation has been very limited, both through the national levels of consultation and at Codex committee meetings and full Codex meetings. Conversely, business representation has been pervasive. In the nineteenth session of Codex, 1989–1991, the vast majority—445 (81%)—of nongovernmental participants on Codex committees represented industry; only 8 (1%) were public interest groups. Twenty-six percent of all participants on Codex committees represented industry interests; in contrast, public interest organizations comprised only 1 percent of total participation. In all, 104 countries participated, as did over 100 of the largest multinational food and agrochemical companies.[38] With such composition and input, the Codex hardly seems a suitable forum to rebuild public trust in the food system and its governance.

Public Accessibility to Decision Making

Clearly, multilevel systems of food governance are in place. Decisions made at the national level have to be harmonized with intergovernmental regulations agreed at the regional level (in the EU) and within international level agreements (such as under the WTO, as underpinned by Codex agreed standards) or crosscutting agreements such as the International Biosafety Protocol. In 2000, an International Biosafety Protocol was agreed under the United Nations Biological Diversity convention, a multilateral environmental agreement covering importation and release of living GMOs in signatory states. The terms of this protocol will need to coexist and complement the decisions arising from the Codex and vice versa, providing another regulatory arena.

The existence of these different levels of governance present differing challenges. These arenas need to be opened to public scrutiny and participation. The needs and concerns of different national publics need to be taken into account and reconciled within a global food system. Democratic processes need to exist to engage the public and generate their trust. The alternative for citizens, as discussed earlier, is to turn to other sources for trustworthy information, such as the NGOs, or for consumers to turn to the retailers.

The implications of these events for the governance of the food system of the European response to large-scale entry of GM foods into their markets suggest a democratic deficit and an absence of public trust in the public institutions. Public governance needs to put the public interest first before commercial interests, at all levels of governance whether local, national, regional, or global. The rebuilding of public trust in our systems of food governance is needed. In particular, this means a reform of the risk analysis process. The perception of risk and the social, cultural, and economic dimensions need to be engaged with and seen as other legitimate factors. Technology must be assessed in a wider democratic process, rather than through narrow, often discipline-confined assessments of risk (such as molecular biology). Such scientific disciplines are largely influenced by social constructions anyway. Thus, the social impacts of new technology and its application, as with GM crops and foods, need to be incorporated into the risk analysis process undertaken by regulators. This should not be left to the market. It is too costly in economic as well as social and political terms. The emphasis of the risk assessment should be based on precaution if for no other reason than that the products of the life sciences cannot be recalled once spread across the farm and beyond.

The controversies over the potential adverse effects of GM crops and foods have covered both environmental and human health (and, to some extent, animal welfare) effects. Nonetheless, there is a tendency for regulators to separate out the environmental impacts of food production from its health impacts.

The European response to GM foods, and the regulatory and supply chain challenges it has prompted, point to the need for a more integrated conception for food policy. The European public is concerned not just about the safety of the final food product, but increasingly about how that food has been produced, its journey from seed to shelf.

Integrating a sound ecological approach to the food system, from production through the point of consumption, alongside a food supply that provides food free from contamination and of the variety necessary for a healthy diet is the way forward.[39] The longer term aim should be a food policy that promotes an integrated model based on a widely defined conception of environmental protection and the promotion of human health, placing that model above the more narrow dictates of commerce and national economic competitiveness.

Chapter Eight

A Societal Role for Assessing the Safety of Bioengineered Foods

Britt Bailey

Our . . . codes of honor, morals and manners, the passionate convictions which so many hundreds of millions share together, the principles of freedom and justice, are far more precious than anything which scientific discoveries could bestow.

—Winston Churchill, 1949

The very scale of the agricultural biotechnology enterprise requires that we examine the full roster of implications of genetically modified foods. Both the volume and the extent of newly introduced genetically altered food crops have ensured they will have a major impact on society, including altering social institutions themselves.[1] When a genetically modified living food product is able to replicate and impose its features on the next generation of consumers, it poses special ethical issues because it can affect generations that have not participated in consenting to its presence in the diet. For this reason, a higher than normal expectation exists that the safety of bioengineered foods be thoroughly assessed. To date, no such assessment has been done. Instead, we are confronted with a "reality on the ground," in which food crops already in place are being examined retroactively for their presumptive safety. In spite of the appeals for caution and greater oversight, 109.2 million acres are currently (circa 2000) cultivated with genetically modified crops worldwide.[2]

New genetic sequences are often introduced into such food crops from disparate species, assuring, at a minimum, that genetically determined factors will cause GM crops to deviate from conventional food products. The central health question, whether such engineered changes are sufficiently biologically novel to pose unanticipated health risks, is only part of a broader ethical inquiry about the aims, goals, and ultimate societal impact of biotechnology.

Uncertainty about Health Impacts

When the introduction of a product or technology has the ability to dislocate an entire food system, including creating structural changes to the foods themselves and to agricultural procedures in general, a broader risk assessment format should be applied. Instead, the more limited regulatory model used for genetically modified organisms is one that has contributed to an adversarial climate. By avoiding a thorough safety analysis, the resulting regulatory vacuum has encouraged clashes between an industry bent on protecting its investments and its products, and vested organizations acting as watchdogs in the name of public health. The resulting dissension has been fueled by a dearth of unbiased information and reporting from industry, coupled with an overpromotion of findings and data about modest and often overblown "risks" from those critiquing the technology.

Had the concerns surrounding genetically engineered foods been addressed by conducting an assessment akin to an environmental impact statement (EIS),[3] we may have avoided not only the anxieties generated by the unintended spread of genetically engineered corn, but also the resulting polarization of positions about the safety of this and related events now held by environmental groups and the biotechnology industry. Lacking a complete premarket review of the risks and benefits of introducing genetically engineered foods, environmental groups have molded the public's concerns around icons that create highly visible but not necessarily central issues, such as the monarch butterflies affected by Bt pollen and Taco Bell corn chips contaminated with Cry9C toxin. The relative lack of proven exigent risks from GMOs may be attributable to this focus on spectacular examples in lieu of a full review of bona fide questions about health and safety.

Regulatory Oversight

The shroud of public uncertainty and misgivings about adverse health effects stems in part from lack of an adequate model of regulatory oversight, which could alleviate concerns and questions surrounding health matters. A further deficiency is the uncoupling of ethical imperatives from a review of genetically modified organisms. An ethics-driven system for regulating and monitoring health effects would be one that focuses on research questions affecting public welfare, weighs benefits and risks, and considers the precautionary principle. Such a system requires transparency of data in order to provide unbiased monitoring and enforcement on the part of governmental oversight agencies. Ideally, such a system would also provide an avenue through which the public could actively participate in decision making.[4]

The argument for public oversight hinges on the reality that genetically engineered foods have infiltrated every major food item sold in grocery stores—without consumer notice or consent. The very inability of consumers to depart from eating foods derived from genetically engineered crops demands supervision and caution in putting these products into the stream of commerce. To do otherwise is to provoke anxiety and to avoid credible management of health issues.

Risks to Susceptible Populations

Currently, no adequate mechanism exists to weigh and evaluate health and safety concerns. The public health paradigm by which we now assess risks to health from genetically engineered foods has served to create skepticism, confusion, and fear rather than provide insight about real risks and benefits. Even more troubling is the absence of bona fide efforts to protect those persons who would be most likely to encounter difficulties from novel food by-products that have made their way into the mainstream food supply.

Were genetically engineered and conventional foods to differ significantly in terms of adverse health impacts, those GMO-based foods could be expected to burden vulnerable populations disproportionately, exacerbating health gaps already created by differential access to health care and exposure to toxic substances. Susceptible populations, including the elderly, women during pregnancy, and those with compromised immune systems, are likely to be most vulnerable to any adverse effects from biologically novel foods. For instance, persons with deficient metabolizing or detoxifying systems might be disproportionately burdened by genetically engineered foods if they contained potentially harmful or allergenic substances or were deficient in essential nutrients. As scientific panels have outlined in their commissioned reports, without sufficient impartial evidence, children will be the first to experience the adverse consequences of the faulty oversight system for genetically engineered food products.[5]

A Protected Industry

In spite of a presidential decree on environmental justice, existing federal oversight bodies tend not to have a differential concern for vulnerable populations in their mission statements. Much of the confusion regarding potential health-related impacts stems from the mistaken belief that our federal agencies are pro-active in their use of justice considerations and protection of the public from potential harms. This belief arises from a misreading of the mission statements of agencies such as the Food and Drug Administration (FDA). While the FDA is directed to protect the public's health, it also plays a gatekeeper role in deciding what foods (and drugs) enter the stream of commerce. In that role, like its sister agency, the U.S. Department of Agriculture (USDA), the FDA must assure that the course of commerce is not impeded by an unnecessarily intense review. Only a tiny fraction of the necessary tests were completed before the FDA declared foods derived via molecular biology to be analogous to conventionally bred foods.[6] This declaration, made in a policy statement in 1992, just a few years prior to the first genetically engineered products finding their way onto supermarket shelves, created a false sense of assurance that bioengineered food products would behave no differentially than do conventional foods in the human body.

Policy Review

But even the seemingly protective policy, whereby any proven deviation from equivalence would seem to disqualify a genetically engineered organism, has proven tenuous. Were it not for the actions of the Alliance for Biointegrity, which challenged the FDA's presumption that genetically engineered foods are nutritionally and physiologically equivalent, the public would assume that the FDA's directives constitute regulations pertaining to bioengineered foods. In rejecting the lawsuit initiated by the Alliance for Biointegrity that challenged the FDA's presumption that genetically engineered foods are nutritionally and physiologically equivalent, the appellate court found that the 1992 policy was nonbinding and did not commit the FDA to any particular enforcement action.[7]

Even before the Alliance's lawsuit, organizations were poring over the policy and pointing out its failings. The court's decision precisely revealed what was already assumed: that the FDA had allowed genetically engineered foods to be marketed without a full assessment of risks or benefits. The lawsuit also revealed critical yet otherwise ignored documents from scientists within the FDA that warned of potential human health effects and urged caution.[8] For example, a memo from the Director of Veterinary Medicine, Gerald Guest, to the FDA's Biotechnology Coordinator, James Maryanski, stated:

> The FDA will be confronted with new plant constituents that could be of a toxicological or environmental concern.
> . . . It has always been our position that the sponsor needs to generate the appropriate scientific information to demonstrate product safety to humans, animals, and the environment.[9]

The Alliance lawsuit exposed a laissez-faire attitude on the part of the government that has only exacerbated public distrust and anxiety. The resulting policy vacuum invites public scrutiny of the agricultural biotechnology conversion. In spite of the fact that 65–75 percent of the foods sold today contain genetically engineered components or by-products derived predominantly from Roundup Ready herbicide-tolerant crops or insect-resistant Bt crops, a detailed review of the scientific literature reveals a systematic dearth of unbiased health safety testing. In spite of such deficiencies, the Life Science Industry has continued to assert its claim that bioengineered foods are safe and substantially equivalent to conventional foods without any independent tests of all the necessary propositions of comparability.[10] Where many organizations have alleged such weakness in health and safety review, the industry has tended to be reflexive and defensive, giving the impression that industry is addressing only what is called to its attention.

Presumptive Safety: Herbicide-Tolerant Crops as a Case in Point

Monsanto has remained the principal developer of herbicide-tolerant crops. In the mid-1990s, Monsanto began flooding the food market with crops engineered to withstand its herbicide Roundup. The genes responsible for producing such effects were obtained from a bacterium, a plant virus, and disparate plants such as the petunia. In 1992–1993, Monsanto submitted data to the FDA to prove that its herbicide-tolerant crops were substantially equivalent to known conventional types.[11,] [12] Their study on Roundup Ready soybeans dismissed the potential human health concerns from ingesting the new genes, asserting it makes no difference in the nutritional value, allergenicity, or taste compared to regular soybeans.[13] In assessing the safety of its Roundup Ready soybeans, Monsanto introduced the altered crop in the feed for animals. Based on the results, Monsanto claimed that supplemental transgenic soybean crops expressing glyphosate tolerance were equal in nutritional value to that of conventional soybeans in animal diets.[14]

While Monsanto's tests on amino acid composition of its engineered soybeans grown in North America showed remarkable concordance to the amino acid makeup of conventional beans, other unpublished studies done on crops grown in Puerto Rico showed variations in the so-called aromatic amino acids such as phenylalanine. The greater abundance of some of these amino acids may result from the enhanced photosynthetic ability that confers resistance to Roundup herbicide.

With a belated acknowledgment from Monsanto that their earlier tests on engineered Roundup Ready soybeans had not been sprayed with Roundup as would normally be required in actual field use, Marc Lappé and I initiated a research study to decipher the nutrient levels (specifically, phytoestrogens) of Roundup Ready soybeans and their conventional counterparts.

Phytoestrogens, predominantly found in soybeans, play critical roles in controlling sexual differentiation, immune function, and mineral availability. Isoflavone-rich soybeans may significantly reduce cholesterol levels and breast cancer risk, but a diet rich in soy has shown to produce uterine adenocarcinomas in adult years when test animals were exposed early in life.[15] The experiments in question exposed four-day-old mice to single injections of the biologically active phytoestrogen genistein in amounts comparable to cancer-causing doses of a proven estrogen-like carcinogen called diethylstilbestrol (DES). The positive cancer results with the plant-derived estrogen are important because DES has been shown to produce cancer in young girls whose mothers were exposed to the carcinogen during pregnancy.

Our initial study suggested a disparity in levels of biologically active phytoestrogens—a 12–14 percent reduction in the Roundup Ready variety.[16] (We have recently reconfirmed our initial finding of below conventional levels in a third variety of Monsanto soybeans). While our results appear to challenge the FDA and industry's claims of substantial equivalence, they also suggest that the lower levels of phytoestrogens found in the engineered soybeans may have health significance. Such a conclusion is buttressed by recent conclusions that some phytoestrogens may support heart function and vascularization, and for that reason, having soy in the diet is "heart healthy." But if the key phytoestrogens are altered in engineered soybeans, this conclusion may be invalid since it relies on testing done only on persons ingesting conventional soy in their diet. At a minimum, we believe our work points to a further need for oversight of genetically engineered products, especially during pregnancy and when such soy-based products may be used to make infant formula. In our view, this pilot work invited further health assessment. Unfortunately, this work has yet to be done.

In its place, the expansion in the soybean market in general, together with an exponential growth in the numbers of acres being planted with genetically engineered soybeans (51 percent of American soybean acres), appeared to prompt industry to defend its products in light of our findings. One week before our results were published, the American Soybean Association launched a Web site denouncing the findings of our study, while Monsanto simultaneously began its own study to examine phytoestrogen levels. Its study found such wide variability in phytoestrogen levels that it claimed it could not conduct a statistical analysis.[17] Nonetheless, Monsanto claimed no differences in phytoestrogens could be found.

From an economic standpoint, industry's advocacy is sensible. In the eyes of an already skeptical and informed public, any discrepant research findings create concerns that lead to public doubt and confusion. Dismissing these concerns is good public relations, even if questionable science. And while the science continues to be debated, the potential for adverse impacts remains a looming question.

The Case of Insect-Resistant Bt Crops

Industry-sponsored studies provided to the regulatory agencies for registration of bacterially engineered insecticide-resistant crops have also been questioned. To date, no one has tested the complete proteins people are exposed to when eating foods containing the by-products of pest-protected crops. In turn, the Environmental Protection Agency (EPA), the governmental oversight agency for insect-resistant crops, has not enforced the required submission of studies on three important safety factors associated with Bt-containing varieties: (1) molecular structure; (2) heat resistance, and (3) size of the molecule. In late October 2001, however, the EPA approved Bt cotton for another seven years without having definitive answers on these safety factors.[18]

Usually dusted or sprayed onto the leaves of growing plants, Bt acts as a larvicide, killing lepidopteran species as they consume their initial food sources. An impediment to consistent use of the natural insecticide is its rapid degradation in sunlight and rain. By incorporating the bacterium into the genome of plants such as corn, potatoes, and cotton, genetic engineers hoped to achieve the desired pesticidal effect while eliminating the pitfall of degradation. In so doing, the biotechnology industry has achieved tremendous commercial success particularly by incorporating specific proteins, such as Cry1Ab, Cry1Ac, and Cry9C, derived from the soil bacterium *Bacillus thuringiensis,* into food and feed crops. It is projected that 20 percent of U.S. corn acreage is currently (circa 2000–2001) planted in Bt varieties.

Food Allergies and Bt Crops

The chief misgiving about the rapid commercial introduction of Bt crops is its potential to cause allergic reactions. Allergens in general have become a serious public health concern, affecting roughly 2.5 to 5.0 million Americans.[19] Up to 2 million, or 8 percent, of children in the United States are affected by food allergies, as are up to 2 percent of adults.[20] Food allergens—those parts of foods that cause allergic reactions—are usually proteins. Many food allergens can still cause reactions even after they are cooked or have resisted the degradation that normally accompanies digestion. Some allergens, such as the Cry9C toxins expressed in some insect-resistant, genetically engineered plants, fit this latter category.

Cry9C and StarLink

Upon review by EPA's scientific advisory panel, the Cry9C protein, coded by genes from the soil bacterium *Bacillus thuringiensis,* was found to have many characteristics [21] of other known allergens, including (1) possible presence in blood after oral feeding (based on animal studies); (2) induction of an immunologic response in animals tested; (3) a molecular weight in the range typical for allergens; (4) resistance to breakdown by acid; (5) resistance to digestion; and (6) its probable status as a glycoprotein. In a recent review, an EPA panel reaffirmed this protein posed a "moderate" allergenic risk.[22]

For these reasons, the Cry9C protein, under the trade name Star-Link, was registered for use *only in animal feed crops.* StarLink, which was developed by Plant Genetic Systems—merging indirectly with Aventis CropScience—could be cultivated, though to do so would require Aventis following an environmental stewardship agreement. The agreement included ensuring farmers were informed of the restrictions imposed by the EPA: the limitation to animal feed and industrial uses, and for farmers to create 660-foot-wide buffer strips around plots of StarLink to discourage genetic contamination via open pollination.

In spite of these safeguards, in early 2000, the Cry9C protein began appearing in hundreds of food products. Nine million bushels of the StarLink variety purportedly had been mixed in silos containing corn used for human food products. The FDA immediately began recalls of products testing positive for the protein, and major food companies such as Taco Bell, Kraft, and Kellogg's began pulling products from their shelves to the tune of millions of dollars in lost revenues.

Hindsight

The initial approval and registration for commercializing StarLink corn came with assurances from regulators and industry that StarLink's Cry9C protein would not find its way into the food supply. Directly after Cry9C's detection, Aventis began petitioning the EPA to approve StarLink for human consumption, claiming its approval would be easier than ridding it from the food system where it was found in copious amounts. The EPA denied this request, and shortly thereafter Aventis voluntarily withdrew its registration for use in the United States.

While it is generally believed that the EPA failed in its oversight by permitting StarLink to be registered for planting only for animal feed, there has been controversy scientifically and philosophically in the outcome of its investigation. The introduction of StarLink corn into the human food supply and its ensuing investigation have followed a pattern of after-the-fact risk assessment that remains the fundamental flaw of our oversight of genetically engineered organisms. Because government agencies failed initially to scrutinize genetically engineered crops, environmental groups have been placed in the position of disputing the bits of science that have been conducted by the industry. Thus industry and environmental groups have been involved in a never-ending tit-for-tat battle over whether one tidbit of data affects the overall safety of a product. This approach has left the public confused. More critically, this reactive science battle has encouraged a neglect of the overall need for a broader consensus of the minimum tests of the safety of a technology that has transformed agriculture as a whole. The current wave of arguing the minutiae overlooks the larger potential health concerns posed to a growing cross-section of the public by GMOs individually and as a group.

Adequately Protecting the Public's Health

While the vigilance of those involved in revealing the deficiencies of scientific testing is commendable, this process is wholly deficient as a socio-scientific model of measuring threats to health. The industry responsible for bringing about this technological revolution has been permitted to continue its product marketing even as it struggles to suppress the prodding of the bioengineering critics. This model will never sufficiently alleviate public health and safety concerns. The public's legitimate concerns regarding genetically engineered foods will not be satisfied until there is adequate enforcement on the part of government agencies charged with protecting public and environmental health. To do so will require reinstituting industry responsibility for conducting full environmental and public health reviews and disclosing their outcomes in the spirit of the National Environmental Policy Act.

Perspective

Factors involved in health do not begin and end with proportionate access to nutrient-rich foods but, of necessity, embrace all risks and benefits. Because the introduction of genetically engineered foods has become a ubiquitous part of the diets of everyone, any assessment of health outcomes needs to be approached from a much larger perspective. It is clear that the governmental oversight agencies have avoided monitoring genetically engineered foods, both in terms of routes of distribution and of any adverse effects. This lack of scrutiny has created a chasm in oversight that has given rise to outcries and objections to biotechnology generally. The failings of our governmental agencies have also created a need on the part of environmental organizations to scrutinize the existing scientific studies to determine and define the need for supplementary tests.

Paradoxically, it is unclear whether any amount of scientific analysis at this point will be enough to convince the general public that genetically engineered foods are safe to eat. What is clear is that no one has determined if genetically engineered foods are physiologically neutral, inferior, or better than conventional varieties already on the market. Even if there could be genuine independent oversight involved in a scientific risk assessment, simply addressing the detailed science fails to consider the moral basis for public claims of under-representation, nonparticipation, and failed consent processes in allowing unlabeled bioengineered foods to flood the market.

Genetic engineering of our food supply has produced a striking degree of social disruption. The full gamut of social questions cannot be answered without understanding agricultural biotechnology's social impacts, psychological effects, and the reliability of long-term safety assurances. Genetic technology has forever altered our understanding of the relationship between science and society. The broad-scale release into the environment of genetically engineered seeds may alter the ecological landscape, threatening traditional weedy varieties and landraces, altering subsistence farming methods, and perpetuating the use of on-farm chemicals at a time when more sustainable methods of farming have just gained a toehold.

Public Involvement

More than ever, the public needs to have greater involvement in the future use and development of these varieties given that the new genetically engineered varieties have virtually supplanted their predecessors. At a minimum, this demand is justified because substantial public funds were used to develop and disseminate the technology. The public should have been more widely involved in decisions as to the technology's use at the outset. While threats such as allergenicity remain genuine and alterations to nutrients could adversely affect (or conceivably improve) public health, the social and environmental perturbations potentially set in motion by widespread agricultural biotechnology are far more alarming. The general skepticism surrounding the acceptance of genetically modified foods is unlikely to be resolved by a single scientific article.

The transforming nature of the technology—now that novel DNA has been inserted into domesticated plant species and is generally irretrievable once it is released via pollen—demands broadly-based public assessment. The development and commercialization of genetically modified foods has fixed in place a method of growing food that many do not believe is or will be sustainable. The greatest failings of the introduction of genetically engineered foods include the disregard, intentionally or inadvertently, for the scale of modification innate to the technology. The questions raised in the name of public health will begin to be addressed only when a full review is conducted, a review that weighs the costs of the risks and benefits, the environmental and social implications, as well as the more detailed health safety hazards.

Given the disrupting potential of genetic modification, it is evident that an assessment of environmental and social impacts should have been conducted at the outset of this technology. This process would have forced greater scrutiny, not just into its health impacts, but into radical and subtle dislocations of farming and social institutions created by agricultural biotechnology at a time in its genesis when public participation would have been meaningful.

Chapter Nine

Learning to Speak Ethics in Technological Debates

Carolyn Raffensperger

This is what you shall do: Love the earth and sun and the animals, despise riches, give alms to everyone that asks, stand up for the stupid and crazy, devote your income and labor to others, hate tyrants, argue not concerning God . . .

—Walt Whitman

Science with a capital S has become our only decision-making tool even though we know there are many other ways to make decisions. Environmentalists have learned to make rock-solid cases for most of their concerns on the basis of science. Global climate change, endocrine disruption, and biotechnology all are subject to the fierce scrutiny of good science.

I have never worried about the ethics of biotechnology. It just seemed like a stupid idea from the vantage point of evolutionary biology. Fish don't sleep with tomatoes. Scorpions don't mate with potatoes. Why did evolution set up barriers between species? What happens systemically when you pull those barriers down? Questions like this offer reason enough for caution.

But a deeper reluctance keeps us from bringing ethics into the discussion when we can just as well invoke science. Ethical questions are rarely welcome in policy circles. We are told to leave our values at the door and make decisions on the basis of science and facts. The Japanese make this argument with regard to commercial whaling. The United States makes this argument with regard to genetic engineering. "Sound science" says little enough so that it may be used to justify almost any action. And most countries devalue ethics in the context of international trade with the sure knowledge that money trumps ethics. Or rather, ethics are seen as an unnecessary trade barrier. This principle is written into many rules of the World Trade Organization, such as the rule that forbids discriminating against products on the basis of how they are made.

While we may be appalled by such policies, we are reluctant to claim moral high ground in condemning them. We, too, would rather base our arguments on science and reason. Most environmentalists are, after all, well schooled in liberal attitudes: Do your own thing as long as your fist stops before it hits my nose. Why should I know better than anyone else?

Furthermore, even if we were free to talk about values and right actions, most environmental activists lack a language to express ethical stances, particularly when it comes to environmental issues. We do not often argue that something is the right thing to do and make a cogent ethical argument. Public health may be the exception. Arguments are grounded in human-rights language—this is at least the beginning of an ethical language. But human-rights language perpetuates the human-centered ethics that many of us pick up from religious and secular culture. Feeding the world means feeding humanity. "Thou shalt not kill," means not killing another human being. "Honor thy father and mother" means Mom and Dad, not Mother Earth or Grandfather Whale.

We all have notions of right and wrong, but it is hard to extrapolate from personal right behavior to generalized ideas about rules or principles for such things as biotechnology. But our very discomfort and ignorance may provide clues about how to establish ethical precepts and behavior that could reorient us in our relationship to our fellow travelers—the other beings on this earth. Perhaps we can learn how to speak an ethical language even in matters of science and technology.

The precautionary principle begins to direct our attention toward ethics in situations where human activities may cause harm but the science is uncertain. A common definition of the precautionary principle was provided by participants at the 1998 Wingspread Conference: "When an activity raises threats of harm to human health or the environment, precautionary measures should be taken even if some cause-and-effect relationships are not fully established scientifically." But it, too, begins with science—or at least scientific uncertainty. It is specific and up-front about the role of science. The ethical component of the precautionary principle is more ambiguous and not mentioned directly except to require precautionary action to prevent harm (see Appendix A).

In many ways the precautionary principle reflects the rabbinic formulation of the Golden Rule. The Christian Golden Rule is usually stated as requiring an affirmative act—"Do unto others what you would have them do unto you." But the precautionary principle, in its attempt to prevent harm, mirrors the Rabbinic Golden Rule as refraining from action: "Refrain from doing unto others what you would not have them do unto you." Of course, the precautionary principle does require action to prevent harm; it is not passive. But rather than doing good, it prevents harm.

Creation Stories

How have we left ethics so far out of the picture? I am relatively well educated, but until recently I have not shared with a broader community the simplest ethical tools for understanding and creating an ethical relationship to the earth.

The debate on genetic engineering has multiple ethical currents running through the slogans but they are seldom examined. For example, "We must feed the world" expresses a human need and implies a corresponding ethic but ignores the costs to the planet of feeding 9 or 10 billion people. Worries about species such as the monarch butterfly are rarely couched in ethical language, although most people respond viscerally to the idea of losing such an important symbol of beauty.

Father Thomas Berry attributes our ethical morass to the false premises of an obsolete story of our origins and our place in the cosmos.[1] Anthropologists note that creation stories are the genesis of our cultural understandings of right and wrong. Accordingly, our incorrect scientific understanding of our creation may indeed have led to our ethical impasse with the earth. Leslie White and others have criticized the Judeo-Christian creation story for similar reasons, but from a different vantage point: The Judeo-Christian tradition says we have dominion over creation and that it is for our use.[2]

David Korten cites other philosophical origins for our difficulties.[3] For instance, Thomas Hobbes, a philosopher of the Scientific Revolution, argued that life has no meaning and human behavior is determined solely by appetites and aversions. Good is that which gives you pleasure and evil is that which causes you pain.

Culture and Ecology

I was wondering about these distinctions when a good friend of mine, a gourmet cook, told me about her college-aged son attending a summer course on folklore and the environment. Each person was asked to bring food that represented his culture. Her son brought a Twinkie. My friend made this deft connection: Their family—white Americans—had descended from kings and queens, so to speak; that is, they had always been the conquerors, never the conquered, so they had never needed to preserve their cultural traditions. They assumed they would prevail, and they had. And she added, "It has come to this."

Assuming that the Twinkie, or another chemically based food, such as Coke, is our representative food, I sought other symbols of our culture. The best I could do was the computer mouse. What was striking was that I could not think of a single part of the natural world that was a symbol of our culture. Not the eagle, not a tree, not a food plant like wheat or potatoes.

Ethics are as much a cultural phenomenon as Twinkies and computer mice. Culture includes knowledge, belief, art, morals, law, custom, and any other capabilities and habits acquired by a person as a member of society, to paraphrase the classic definition by E. B. Tylor.[4] Culture is the patterned behavior learned by each individual from the day of birth as s/he is educated by parents and peers to become, and remain, a member of the group.[5]

As an anthropologist I am fascinated by the cultural norms expressed in the list written by Robert Fulghum titled "All I Really Need to Know I Learned in Kindergarten."[6] Many of these are ethical—sharing, playing fair, not hitting, cleaning up your own messes, saying you're sorry—and we really do learn these things in our earliest socializing experiences.

Korten sums it up this way: "Culture is the sphere in which the society defines the values, symbols, and beliefs that are its sources of meaning and identity. It holds the normative power to determine what is valued and to legitimate institutions and the uses of the power, resources of polity and economy. Though cultural power may seem weak compared to the powers of coercion and exclusion, it is ultimately the decisive power in any society, as it is the foundation on which all else rests, including the powers vested in the formal institutions of the polity and the economy."[7]

A subset of anthropology—the study of culture—is ecological anthropology. It is the study of how culture articulates and regulates the human community with the natural world. Many cultures that produce or hunt and gather their own food have intense cultural norms about what is good and right in human environmental relationships. These cultural practices prevent the development and spread of disease. They guarantee not only full harvests but also future harvests. People like the Hopi or tribes in New Guinea understand that their actions—killing a snake, having too many pigs, or eating fruit before the harvest festival— have vast consequences for their survival. Ethics and spirituality guide behavior, especially in relation to the natural world.

We no longer have cultural guides for our relationship to our fellow beings. Our culture regulates our behavior to other humans and to technology, not the natural world. Accordingly, the only tools our culture provides for relating to the earth are scientific and technological, not ethical.

Getting to Ethics

What do I mean by ethical tools and language? The *Encyclopedia of Philosophy* says that the term *ethics* is used in three different but related ways, signifying (1) a general pattern or way of life, (2) a set of rules of conduct or moral code, and (3) inquiry about ways of life and rules of conduct.[8] What we lack as a culture is both a general pattern or way of life and a set of rules of conduct or moral code that sets out a right relationship to the earth. We have an abundance of academic investigations into ethics but little that has filtered to the populace at large.

Is this a hopeless situation? Perhaps it is in the short term. But there are some things we can do to establish ethical guidelines within our culture that will reorient us in the natural world.

Reason and Science have gotten us into boxes. From within those narrow perspectives we can't undo hundreds of years of bad theology, bad science, and bad economics—much less come up with elegant, clear, ethical environmental guidelines. We are going to have to start someplace else.

We-Thou

We do have some cultural fragments that might help us weave together a whole cultural cloth of an environmental ethic. Some of the fragments are present in Fulghum's list. I want to suggest two large ethical ideas that extend these cultural ideas to the natural world.

First, it is imperative that we find, recognize, and honor people who have the gifts for living in a right relationship to the earth and who can train others to do so. Some people within almost every culture have special responsibilities to serve as mediators between the natural world and the human community. This is a set of skills, which is almost extinct in

Western society. Shamans, scientists, artists, farmers, and doctors all have had skills in understanding and mediating this relationship. Doctors seek to understand the microbes that cause disease and heal the ill. Farmers have enlisted the life-giving forces of soil, water, and seeds to provide food for humans and other species. Shamans, scientists, and artists help people in their local environments see, and therefore relate, differently to the natural world.

We need to find those among us who have a gift to mediate the relationship between humans and the rest of the natural world. To paraphrase the twentieth-century Jewish theologian Martin Buber, people with this gift help create a "we-thou" relationship rather than a "we-it" relationship with the other beings.

Respect

The second notion ties into the first. A we-thou relationship is predicated on respect, not reason. My colleague Anita Bernstein has proposed that the Respectful Person standard replace the Reasonable Person standard for matters of law, especially offenses against dignity. That is, the standard by which behavior should be measured is not whether someone might have reason to engage in it, but whether such behavior embodies respect for others.[9] The earth deserves our respect. Reasoning about how much poison and damage the earth can tolerate, and why such damage is justified, is only going to perpetuate the destruction and hasten our own demise. We should judge our actions by how respectful we have been.

I suspect that Adam Smith, a moral philosopher in spite of his forays into economics, might agree. He wrote in *The Theory of Moral Sentiments* that sympathy is the mainstay of morality. Respect and sympathy are analogous experiences. Both are expressions of fellow feeling. Genuine morality toward other human beings goes deeper than rules of right and wrong, just as ethical behavior toward the environment must have a deeper basis than science and fact.

Perhaps the people most able to express an environmental ethic predicated on respect or sympathy are artists in all media. They can convey the wonder, majesty, humor, and beauty of the earth and our fellow beings.

Black Popcorn, Prayer, and Japanese Research Animals

At some point in my professional work on the precautionary principle and biotechnology, I realized that I knew more about biotechnology than I did about conventional seed breeding. A few years ago, I went to hear a beloved friend and Master Gardener, David Podoll, give a lecture on seed breeding and seed saving. My friends in North Dakota had chided me for not saving seeds, but it had seemed too complicated and difficult. My garden is huge, lush, gorgeous, and teeming with all manner of fruits, vegetables, and flowers. I write a cooking column for a sustainable agriculture newsletter. Wasn't that enough? Perhaps it was. Or at least this is what I thought until that lovely day when I heard David speak. I learned a lot about the effects of latitude on garlic, safe sex for squash, and how temperature can change the color of a seed.

But I went home that afternoon and harvested black popcorn from seeds that David had given me months earlier. I was electrified with excitement. The cobs had so much variability and so much beauty. Out of hundreds of cobs there was one white cob with a few lavender kernels. I am now breeding black popcorn, buttercup squash, oats, and Hutterite soup beans. I confess to being in love with these plants. They are my companions and teachers. I think we are establishing a relationship of respect, although it is different from the respect and relationship I have with wild things. These seed have given me a glimpse into an ethic of respect that may encompass the plant kingdom.

There are wonderful examples of humans acknowledging their connection to animals. For instance, Barry Lopez describes his encounters with road-killed animals. He gets out of his car and bows out of respect and sorrow for the loss of life. Similarly, I heard recently that Japanese scientists have a ceremony of respect for the animals that are used in research. They express gratitude for the sacrifice the animals are making. Wise move.

Earth Family Values

What would a relationship of respect, a we-thou relationship, look like with all the plants and animals that we depend on? How can we fashion an environmental ethic from the whole cloth of culture—even Twinkie-eating, technology-fascinated American culture?

I thought that posed a difficult, maybe impossible challenge. Nevertheless the Science and Environmental Health Network staff decided to call together some of the wisest people we know to think about these questions. In a late fall 2000 gathering in the Adirondacks, two dozen ethicists, philosophers, scientists, writers, and environmentalists took on the challenge. As it turned out, they reached right down to basic values that reside in human nature and thrive among us, even if our culture barely acknowledges them. Together, we selected the values we thought were particularly important right now, if humans are to survive on this earth. We agreed with Aldo Leopold that these values directed us to extend our very notion of family and community to include the earth and other living things. Without much prompting, the group quickly identified respect and sympathy as essential values, but we went on from there to draw up a more comprehensive list. We couched these values in an affirmation of the place of humans in the ecosystem.

The Blue Mountain Lake Statement of Essential Values

Values become actions. Too many of our actions are killing our planet, our communities, and our spirit. Our actions are killing our loved ones. We are diminishing the future for everyone and everything.

Particular values form the basis of our survival. When practiced, they help us live in reciprocity with nature and with each other. We are the relationships we share, and we are permeable—physically, emotionally, spiritually—to our surroundings. Therefore, we hold these values as essential:

- Gratitude, because our lives depend on air, water, soil, plants, humans, and other animals;
- Empathy, because we are connected with all of creation;
- Sympathy, because we all experience suffering and death, both necessarily in the course of life and unnecessarily when these values are not practiced;
- Compassion, because it moves us to attend to suffering and injustice; and
- Humility, because we cannot know all of the consequences of our actions.

We belong to the community of the Earth. It is the source of our own life, and our actions affect its well-being. Therefore, we practice:

- Respect, because it is fundamental to good relationships;
- Simplicity, because we are only one species sharing Earth with many others;
- Humor, because life is good, and humor disrobes tyranny and absurdity.

Human beings need sustaining social and natural environments. No one by law or habit is entitled to rob others or future generations of a diverse world vibrant with hope and possibilities. We have an obligation to restore social and ecological fabrics that have been torn by violence or exploitation.

We affirm that all being is sacred and has intrinsic value that is not monetary. People who hold these values outnumber those who do not. We draw strength from each other. As we abandon harmful activities, we take nature as our guide. We explicitly consider the effects of actions on individuals, families, communities, species, landscapes, regions, and future generations.

It is through love for the particular—a child, a neighborhood, a family of otters, and a meandering river—that we find our way to a sustaining relationship with the Earth and our communities.

Blue Mountain Center, Blue Mountain Lake, NY, November 12, 2000[10]

Ethics and Biotechnology

I still believe that agricultural biotechnology makes no sense for biological reasons—fish don't sleep with tomatoes, and so forth. But the values enunciated above provide ethical reasons on which to hang skepticism as well.

A little empathy would help to stretch our powers of observation to see the real and potential victims of biotechnology: the monarch butterfly caterpillars that die because they eat wayward pollen, each cell of which contains part of a microbe meant to kill corn pests; organic farmers who cannot protect their crops from the same pollen, and who may be sued by biotechnology companies if they save and replant their tainted seeds; wild fish populations that may be wiped out by engineered superfish escaping from tank farms.

Humility would stand us in good stead, because we do not know all the consequences of engineering species. There are, of course, apparent benefits of biotechnology, and apparent beneficiaries. But we do not know enough even to understand whether those benefits are true or false, or whether they will lead to greater problems.

We should be suspicious by now of unidimensional, technological answers to such complex challenges as "feeding the world." A respectful stance would begin with the complexity of both culture and nature for clues to how that should happen. The chances are small that the profit-making enterprise that has engendered and arisen around biotechnology will feed the world. The chances are great that, like the Green Revolution that preceded it, the biotechnology revolution will end up enslaving economically as many as it frees from hunger. The chances are even greater that if we were more attentive to both the cultural and the physical soil in which crops are grown, we would learn how to feed ourselves in ways that would sustain the earth as well as the human population.

The restraint we must practice is restraint from the hubris that has gotten us into such technological messes time after time. Respect, humility, restraint, even humor—that we have to learn all these lessons time and again—are as important as scientific knowledge and investigation and our boundless technological creativity. We cannot afford to leave values such as these at the door when talking about biotechnology.

Chapter Ten

A Perspective on
Anti-Biotechnology Convictions

Marc Lappé

The evolution of dissent against genetically modified organisms illuminates the special vulnerabilities of the arguments used both to support and to oppose agricultural biotechnology. At the core of the debate between anti-biotechnology activists and its proponents is the assertion that no meaningful differences exist between conventional and genetically engineered foods. Establishing the truth of this assertion was critical to deregulating various commodity crops.

Substantial Equivalence

Historically, the Food and Drug Administration (FDA) used the regulatory concept of "substantial equivalence" to permit the successive incorporation of new varieties of conventional foodstuffs without encumbering the system with unnecessary bureaucratic rules. A Rome apple can replace a Pippin variety; a new Yukon Gold potato an Idaho standard, and so on, without any new testing. Under these regulations, the FDA permits any new food crop to be registered under existing guidelines for conventional food as long as it physically resembles its predecessor and is not declared by the producer to contain a new additive.[1] This system was clearly designed to assure the easy transfer of new products under old banners where the novelty at issue was created by minor, naturally occurring genetic mutations in the plant cultivar within the species in question. But the premise of this old FDA rule is challenged where genes are introduced across species lines.

The industry believes genetically modified products should qualify under the FDA rules for a blanket exemption because these products are not "substantially different" from their conventional counterparts. Such a viewpoint invites closer scrutiny. The core assertion of equivalency by the agricultural biotechnology industry rests on the belief that genetically engineered crops are not different simply because they are engineered. Proponents of this view—that it is the product and not the process that matters—contend that genetically engineered foods should be judged only by their properties, not on the process by which they were produced. By this token, corporate spokespersons asserted that genetically modified organisms (GMOs) are just as safe as their predecessors.

But, as Dr. Louis J. Pribyl (one of the seventeen government scientists who worked on a policy for the introduction of GMOs in the early 1990s) pointed out, merely asserting equivalency and safety is insufficient to permit the widespread application of genetic technologies in agriculture. Pribyl had seen research studies that concluded toxins could be inadvertently created when new genes were put into plant cells. In a memo to his superiors at the FDA, he addressed the issue of unanticipated risks: "This is the industry's pet idea," he wrote, "that there are not unintended effects that will raise the FDA's level of concern. But time and time again, there is no data to back up their contention."[2]

Pribyl's concerns point to an issue of the moral and scientific legitimacy of industry's assertion of equivalency. Under the Aristotelian principle "to treat like things alike" is its corollary: to treat different things differently. Without sound data proving equivalency, treating GMOs as if they were exactly the same in form, content, and flavor as conventional food crops is both philosophically and scientifically flawed. More to the point, to equate a novel, genetically engineered plant with its predecessors is unscientific in the absence of definitive testing. Commonly, the genomes of engineered plants have not only been sullied by one or more copies of the introduced gene cassettes that are randomly inserted into their chromosomes, but have undergone field selection that assures only that they physically resemble a typical soybean, corn, or rice plant. The review done by our team at the Center for Ethics and Toxics (CETOS) of the very first field tests of GMO soybeans, for instance, revealed that many of the identically altered plants were either too rangy, too squat and bushy, or simply had abnormal leaf placement to qualify as "identical" to conventional soy plants. Yet, the ones that were chosen to propagate the original Roundup Ready lines had no more guarantee of *actual* equivalency than would a child's choice of one red block for another.

In 1992–1994, the industry countered these thoughts by pointing to a small group of field tests, feeding studies, and compositional analyses that it interpreted as proving equivalency. The FDA and the Environmental Protection Agency (EPA) accepted these tests as the final word. For example, Monsanto conducted an initial battery of tests in the early 1990s in which genetically modified soybeans were "safely" fed to catfish, cattle, chicken, and laboratory animals with none or only "minimal" changes in protein and milk compositions.[3]

In their research, Monsanto scientists asked only the broadest questions relevant to equivalency: Do GMO-fed animals thrive and grow at the same rates as their conventionally fed controls? Do they show any obvious signs of disease or disability? Is their behavior normal? If no detectable differences were apparent, they expected (and got) regulatory agencies to give them a green light. In 1994, the resulting regulatory largesse hastened the introduction of at least one major genetically engineered commodity, Roundup Ready soybeans, into the marketplace.

A Closer Look

Based on these initial batteries of tests, Monsanto scientists concluded that all data indicated their genetically modified products were strictly equivalent to their conventional varieties. In fact, Monsanto trumpeted this "fact" in the title of the key piece that presented its data publicly: "The composition of glyphosate-tolerant soybeans is equivalent to conventional soybeans."[4] On closer inspection of the underlying data (which was taken from files obtained from the University of Arkansas Extension Service, and represented the first full year of field testing of comparable Roundup Ready and conventional soybeans grown side by side), our team at CETOS was the first to report (in Lappé and Bailey, *Against the Grain*, Common Courage, 1998) that yields from engineered soybeans were generally inferior to conventional crops in the field; and that Monsanto failed to treat its laboratory soybeans as they would be treated in the field—no spraying of Roundup was done on the genetically modified soybeans designed to withstand such treatment. When Monsanto scientists field tested their soybean as an animal feed supplement, they found significant differences—notably, catfish protein was higher in test- versus control-fed fish, and dairy cattle on GMO-supplemented feed had higher fat content in their milk; and as our own work at CETOS was to show, subtle differences in phytoestrogens separated GMO and conventional soybean varieties.[5] Subtle differences in some amino acids were also present in soybean test samples from Puerto Rico—but were not included in Monsanto's published findings.

On the strength of these clues, Britt Bailey and I repeated the Monsanto work with sprayed conventional and Roundup Ready soybeans. Our work on sprayed soybeans showed the Roundup Ready soy had a consistent deficiency in at least two phytoestrogens (genistin and daidzin) when compared to their genetically equivalent counterparts. Each of three pairs of Roundup Ready and conventional soybeans were eventually tested, including two from Hartz seed and one from a subsidiary. Each comparison showed that two, and only two, phytoestrogens—genistin and daidzin—were diminished in the Roundup Ready varieties. Both of the phytoestrogens affected were the most important in terms of their biological activity as plant estrogens. Because at least one of these phytoestrogens (genistin) has since been determined by the National Toxicology Program (NTP) to be a reproductive toxicant, we believed our work had clinical relevance. Monsanto took the position that such "minor" exceptions were simply quibbling. Monsanto observed that natural variation in phytoestrogens among soy plants could be as much as 100–300 percent of normal because of different growing conditions, and that therefore our results (admittedly, only a fraction of these values) could be ignored. We believed our comparisons were important and valid because we compared Roundup Ready and conventional crops grown under nearly identical conditions.

I believe findings, such as ours, that demonstrate scientifically significant inconsistencies in food composition challenge the premise of absolute equivalence. But when we sent our data to the FDA, we received no response. The FDA appeared to agree with Monsanto, and reasserted its contention that in spite of our findings, GMOs in general, and soybeans specifically were still "substantially equivalent" to preexisting foodstuffs, and thereby required no special regulatory action.[6]

In contrast to the relative inaction of the U.S. FDA, the European Union has taken the global lead in regulating GMOs: In 1990, the European Economic Community (EEC) issued edicts that required contained use of GMOs and proscribed the deliberate release of experimental GMOs.[7] In 1997, the EEC issued labeling requirements that became mandatory in 1998. And in 2000, they established a 1 percent threshold for GMO products or contaminants in non-GMO food and created a new authority to review the incorporation of GMOs into food additives and flavorings (EU Regulation 49/2000).[8] Most critically, the EU does not consider GM fresh plant foods as substantially equivalent to their predecessors and requires that all GM foods introduced into the marketplace be labeled as such (EU Regulation 50/2000). The historical roots of this discrepant response between the United States and countries abroad are revealing.

Historical Considerations

Although the groundswell of grassroots opposition to GMOs in the United States did not gather momentum until the late 1990s, the basic technology and initial testing of GMOs was begun almost a decade earlier. By 1986, new applications of recombinant DNA technology had already been tried in laboratory settings. Concordantly with their technical support teams who were planning the first field testing of GMOs, corporate executives attempted to assure their products would be able to meet any regulatory hurdles created by government. Late in 1986, four key executives from the Monsanto Company made an administrative decision: They would go to Washington to encourage the federal government to institute regulations over their own industry. According to a *New York Times* report, the move was predicated on a company strategy to preempt the regulatory process by appearing to initiate self-regulation. Under such circumstances, the executives reasoned, they could expect the government to adopt rules favored by the industry.[9]

Inexplicably, the corporate executives found their efforts largely negated by the inaction of regulatory officials. A review of documents available during the period revealed no real impetus from Washington to control the burgeoning biotechnology industry. In such a soft regulatory climate, Monsanto's team abandoned their effort for self-directed controls, perhaps believing that no regulations were better than even the friendliest ones. One crop after another was "deregulated." From the regulatory perspective, such a move was justified based on the industry's assertion of the general equivalence and acceptability of engineered foods. Under the ensuing USDA and FDA regulations, GMO crops needed only to comply with the general rules of food hygiene and pesticide levels that applied previously to conventional types.

In the end, it appeared as if industry effectively wrote its own regulations. But the industry was unprepared for the social and political upheaval generated by its success in rapid commercialization of GMO food crops. In the mid-1990s, as Monsanto and other large corporations such as DuPont and Rhone Poulenc began to enter the fray, major issues of corporate control of farming, including the displacement of small farmers as a result of the increasing scale of operations became "hot" issues within the environmental community. By 1998, Monsanto had taken over much of the U.S. corn, cotton, and soybean seed market, raising the question of possible monopolistic control of food crop germ plasm. Had Monsanto continued its aggressive seed producer acquisitions, and not divested itself of some holdings, it would clearly have become a monopoly.

To "protect their products," industry representatives aggressively attempted to quell the voices of potential critics. For instance, Monsanto allegedly presented the Fox TV Network, who employed Florida reporters Steve Wilson and Jane Akre, with "evidence" that Wilson and Akre were biased and that their reporting and the resulting series would unfairly disparage a Monsanto product (bovine growth hormone–stimulated milk) were the series permitted to air. Monsanto allegedly demanded that the reporters be denied air time or that they modify their report. The station acquiesced to this demand. The reporters were fired, but later sued and Akre won a verdict that gave her reimbursement for court costs, expenses, and legal fees.

Britt Bailey and I, editors of *Engineering the Farm*, were also attacked for an initial assessment of Roundup Ready technologies, and our publisher was threatened with a lawsuit if he proceeded with publication of *Against the Grain*. As a result, the original publisher dropped the book and had to destroy part or all of an initial press run. Eventually, the book was picked up and published by another publisher (Common Courage Press, 1998), although the book was effectively suppressed for almost a year. No formal legal action by Monsanto was ever taken.

Consumer Revolt

By far, the strongest consumer reaction against GMO producers has been against the biotechnology food industry's reluctance to require labeling of its products. I experienced this "reluctance" firsthand in the wave of coordinated opposition to bill SB 1513, which would mandate labeling of GMO-containing food products, and which I wrote with California Senator Tom Hayden in the 2000 session of the California Legislature. This bill reasserted the public's right to clear notice of food composition and would have required a disclosure of any product whose components included 1 percent or more GMO ingredients. The bill exempted restaurants from the labeling requirement and only pertained to those GMOs that actually contained altered gene products. (Similar rules were already widely adopted by the European Union.) But more than twenty different organizations, from the California Grocer's Association to the Farm Bureau, actively lobbied against this legislation. By thwarting this and related legislation in at least nineteen states, the biotechnology industry has so far quashed labeling legislation in the United States. In so doing, it also prepared the soil for an upsurge of consumer discontent.

Disclosure and Compromise

To quell opposition to national and global concerns, on January 18, 2001, the FDA issued new rules that would allow foods to be labeled voluntarily as being biotech-free or with the phrase "contains genetically modified ingredients." These rules were issued in two documents as part of a joint proposed regulation: "Premarket Notice concerning Bioengineered Foods" and "Guidance for Industry: Voluntary Labeling Indicating Whether Foods Have or Have Not Been Developed Using Bioengineering." While industry perceived the new labeling rules as meeting consumer demands, activists protested that they fell far short of the mandatory labeling requested.

As part of the new rules, the industry would be mandated to "consult" with FDA by notifying them of new biotechnology products at least four months before their marketing rather than conduct what was before a voluntary consultation for pre-market review of their products. This rule required only that the marketer include a description of the food, its method of development, the use (if any) of antibiotic-resistant marker genes, any allergens that might be present, and a basic comparison to conventionally developed foods of a similar kind. But the activist community pointed out that the FDA did not require either form of labeling, or mandate pre-market safety testing. In the view of Kimberly Wilson of Greenpeace, such a decision continues to deny consumers "free choice in the grocery store."[10]

Transparency

The concept of transparency and openness remains a key ethical issue to be addressed. Current FDA guidelines permit the producer of a genetically engineered food to claim that information given to the FDA is a trade secret and hence may not be revealed to a consumer. Further, the producer of an engineered food crop is exempt from the National Environmental Protection Act, a sweeping omission that would exempt a foodstuff from being examined for its environmental impacts in general, as well as its human health impacts. Most critical, by not requiring a pre-market or an environmental review, nothing in the mandate's requirements requires the manufacturer to provide a mechanism for tracking its foodstuffs, should they escape controls (as did the StarLink corn). While the FDA did ask for public comment on whether or not a pre-market notification letter was needed to include methods for tracking and/or detecting its food once it is in the marketplace, no such requirement was added to the final rule. This remarkably passive rule was issued in spite of strong concerns about safety and related issues.

The National Research Council Studies

In 1999, the FDA encouraged the National Academy of Science (NAS) to conduct a third report on potential health risks posed by one of the most pervasive and potentially risk-laden products, plants that contain their own pesticides in the form of gene products from *Bacillus thuringiensis* (Bt). The National Research Council (NRC) tackled only the question of demonstrable risks of these pest-protected products, not the method of their production or its likely contamination by related organisms. In this sense, the NRC report fell short of asking the broader questions of novel risks posed by GMOs more generally—and if such exist, how they are to be assessed and quantified.

The NRC found that at least one safety issue remained unresolved: the allergenicity of Bt-containing crops and the current insufficient testing to determine the immunologic effects. As such, the NRC recommended a battery of tests be required prior to marketing; these testing protocols have yet to be adopted. Asking only if the new foodstuffs are "safe" from conventional perspectives also failed to acknowledge that any self-perpetuating differences in GMOs might create novel concerns for the future of agriculture in general, and the stability and preservation of seed germ plasm in particular. For instance, by avoiding the issue of cross-pollination and gene contamination, the NRC may have missed the agricultural problem of the century: the perversion and contamination of historic seed stock with incompletely understood and novel germ plasm from Bt-engineered crops such as corn and cotton. (In late 2001, gene products from both crops in Mexico and in India respectively, were reported to have cross-pollinated and spread into the countryside to contaminate native species.)

Even on the central issue of the report itself—namely, the loopholes in immunologic testing—the NRC can be faulted. The issue of allergenicity cannot be anticipated with the NRC's proposed test regime, which is largely limited to animal models. As shown by the current dilemma of finding individuals with alleged allergic reactions to a new corn antigen without having a test kit available for the presence of IgE antibody, the NRC did not anticipate any of their concerns being "real" health problems. Deciding if an allergic reaction has occurred is not resolved by spot-testing a few persons for IgE antibodies to a new antigen, or performing a perfunctory set of immunological tests in animals.

As a consultant to the NRC, I was asked to limit my scrutiny to scientific data and to focus solely on scientific questions of risk while eschewing political, social, or philosophical issues. The NRC admitted that, by intention, its study "does not address philosophical and social issues surrounding the use of genetic engineering in agriculture, food labeling, or international trade in genetically modified plants."[11] The NRC's scientific staff and consultants could be faulted for not looking further into precisely these moral, ethical, and political issues raised by the burgeoning field of genetic engineering. The NRC acquiesced to the limitations of its study, noting that its review was limited by its assignment.

Most immunologists would probably concur we have a very incomplete panel of toxicity tests to anticipate protein novelty. If true, this failing suggests how unprepared we are for monitoring the public for any adverse reactions to GMOs. When novel proteins enter our biological universe, it may be months or years before full-blown allergies develop and are recorded by regulatory agencies. We may well miss such events, as tracking methods are flawed and our ability to recreate dietary exposure is hampered by a lack of labeling. What is needed at the outset is a set of concerns that recognize genetically modified crops for what they present to the consumer: the social and health implications of protein novelty in food.

With hindsight, NRC's emphasis failed on three counts: (1) It precluded consideration of genuinely pro-active health measures, based for instance on the precautionary principle; (2) it denied the validity of ethical, moral, or religious concerns about the nature of food and its production; and (3) it preempted the debate about the overall acceptability of genetically engineered foods. By avoiding the major ethical (and moral) issue of whether or not genetic modifications comprise a fundamental shift in technology per se, and by not addressing the duties involved in pre-market testing and review, the NRC asked for a review that had a foreordained answer: Insufficient evidence exists on which to base health concerns from GMOs. While the NRC did recommend additional safety testing, such as allergenicity testing, its report fell far short of providing the scaffolding needed for the systematic review and orderly introduction of novel agricultural products.

Despite such third-party concerns, Aventis, Searle, Monsanto, and Dow/Pioneer have continued to market new plant-incorporated insecticides based on the Bt technologies that they contend will further reduce pesticide reliance for control of the corn borer and other pests. Such chemical reliance now includes nearly 96 million pounds of pesticides used against corn pests yearly, including dangerous organophosphates that industry spokespersons claim (without as yet adequate documentation) are being phased out because of the advent of pest-protected crops.[12] Were water-contaminating and potentially carcinogenic herbicides such as those in the atrazine family that are widely used on corn actually reduced, it would have a salubrious effect on the ecosystem. But little or no GMO-related change in herbicide or insecticide use has occurred with the exception of pest-protected cotton.

The actual reductions in pesticide use appear modest. If Bt proteins truly reduced the reliance on chemical pesticides, that "fact" should be demonstrable with actual statistics. For at least the first three years of incorporation of Bt technology into corn, the data have proven ambiguous, suggesting a decrease in pesticide reliance in only one of the first three years during which mass operation of the Bt program was underway. Overall, through 1998 the Department of Agriculture found only a 1 percent decrease in the amount of pesticides used on corn, cotton, and soybeans.[13] Most of that reduction, however, was in use on cotton crops. Based on a review by agriculture consultant Charles Benbrook, published in the October 2001 issue of *Pesticide Outlook,* overall pesticide use has actually increased slightly since the introduction of pest-protected, genetically engineered crops. While subsequent reductions in cotton have proven more significant, modified plant genetics is only part of the reason for pesticide use reduction. The European corn borer is not a major problem currently, and cotton-associated pests such as lygus and mites have put little pressure on growers in the 2001 season, resulting in a much lower pesticide reliance independent of Bt-protected cotton.[14]

Proponents of genetically engineered food products may be correct in arguing that so-called medical and health-related risks of new varieties remain largely hypothetical. As I have shown, they can point to studies that appear to establish that the new product is for all intents and purposes identical to the old one. They also argue that any newly added gene products are generally digestible, nutritiously equivalent, and therefore allegedly possess "no allergic concerns" when compared to conventional varieties.[15] But this stance belies existing uncertainties, undermining the importance of assuring safety in advance of marketing, a key provision of the precautionary principle.

Feeding the World

As a counter to the groundswell of opposition, the Biotechnology Industry Organization (BIO) accused the activist community of thwarting the effective introduction of biotechnology-derived foods to the developing world, where they asserted such food was urgently needed. BIO's consultants pointed out that food experts estimate it will be necessary to double food production in less than two generations to meet the burgeoning world population; some 2 billion more people will likely be born in the developing countries alone by 2020. Thus, while this concern may be bona fide, it did not appear to be the original motivation of the biotechnology industry's campaign to convert the American (and ultimately, the world) food supply. Nor is it likely that biotech-generated food, which has yet to achieve the economics of scale achieved by the first Green Revolution, is any more likely to feed the world than the present food supply unless problems of distribution and access are solved. I note that the Food and Agriculture Organization (FAO) believes existing levels of production ensure adequate food through 2030.

Based on all the publicly available information we could access, it was not until the end of the 1990s that any of the corporations developing GMO products formally asserted that they had developed their technologies to meet the need for cheap and/or abundant food. Looking back, it is evident that assuring adequate food supplies or consumer satisfaction was never a primary part of Monsanto's (or any other company's) plan. Almost all of Monsanto's initial products were designed to make food crops "friendly" to the farmer, not to the consumer.

Ostensibly, "technology fees" and contracts that required the farmer to rely solely on the chemicals produced by the seed maker's parent company were merely ways to return the capital outlays made by the sponsor in developing its product. But as 1998 led to 1999 and 2000 to 2001, with continued increases in this fee, it began to look—as with turnpike and bridge tolls—as if the farmer and the consumer were being asked to pay for a commodity long after its start-up costs were recovered. With the notable exception of the late-entry vitamin A–enriched Golden rice, we are aware of no GMO created intentionally to increase human nutritional value.[16]

Ironically, the plethora of potentially beneficial products from agricultural biotechnology—more digestible grains, better cassava varieties, vitamin A–enhanced Golden rice, and pharmaceutical-containing products such as bananas with hepatitis B vaccine—appear genuinely beneficial though remain incompletely commercialized to date. Other advances like phytase-containing grains did increase animal feed efficiency and, in the case of blight-resistant Hawaiian papaya, salvaged a disease threatened industry. But these innovations appear to be exceptions to the general pattern of mass planting of staple food and forage crops.

Prospects and Promises

The current BIO position that biotechnology overall is delivering on its promise of providing more nutritious, cheaper, less pesticide-intensive crops remains a questionable assertion. Eventually, the argument goes, these benefits will trickle down to people in the developing countries. BIO can point to experts from Third World countries, such as Dr. Violeta Villegas of the Philippines. Villegas has argued that even today's biotechnology would help reduce her country's agricultural and medical problems, and should be encouraged to expand in spite of the popular revulsion evident in countries such as India and Brazil.[17]

The counterpoise from the activist community is that biotechnology has neither fulfilled its promise to the developing world, nor developed a strategy for consumer involvement with clear public health and ecological safety guidelines in place. Without these guidelines, such as the recently approved Biosafety Protocol, the promise of abundance through biotechnology may be a mirage.

A Moral Vocabulary for GMOs

If nothing else, the biotechnology revolution in agriculture has demonstrated the need for a new moral vocabulary to address how farming should go about meeting the nutritional and political needs of the world. Does an industry that is developing products designed to replace existing agricultural systems have a duty to ensure those products meet sociopolitical as well as nutritional or economic needs? Or are the obligations of a new agricultural order no more or no less than those of its predecessors?

To date, the debate has tended to miss these broader questions. As geneticist Richard Lewontin has pointed out, opponents of GMOs have historically addressed five general issues concerning GMOs: (1) threats to human health; (2) disruption of the environment and the ecosystem; (3) creation of novel and potentially disruptive new pests; (4) dislocations of agricultural production, especially in the Third World; and (5) objection "on principle" to the method of production of new crop plants and species.[18] Additional questions, including the unilateral decision to introduce genes across species lines and the absence of a precautionary principle guiding safety testing, raise more fundamental problems concerning the role of science and the exercise of corporate power.

Problems and Deficiencies

Outside the United States, in places such as Great Britain, Canada, and the European Union, review agencies encourage awareness of these broader questions posed by corporate agricultural biotechnology. Some of these agencies have found major deficiencies in the industry's risk assessment process. In particular, the United Kingdom's renowned Medical Research Council has spelled out the inadequacies in the present risk assessment superstructure: No one has thoroughly assessed any of the following possibilities: (1) the altered nutritional quality of genetically engineered foods, (2) their acute or chronic effects, (3) their impact on the immune system (including its normal development), (4) the transfer of antibiotic-resistant genes to human pathogens, or (5) the inadvertent creation of novel infectious disease risks such as might occur through the use of hazardous viral gene vectors.[19] Perhaps more critical, no one has assessed the overall stability of the newly created genotypes with the crude, first-generation genetic techniques currently in use.

Tracking Needs and the StarLink Episode

The biotechnology industry has played down the most serious genetic contamination episode to date—the ostensibly inadvertent spread of a strain of corn (StarLink) that contains an unapproved and potentially allergenic protein known as Cry9C. In early 2001, Taco Bell products were found to be contaminated with Bt-corn approved only for livestock feed. Over the next six months, over 180 consumer products were found to contain this contaminant.

The sudden appearance of this marker points to the ease with which GMO proteins can spread—and, ironically, to the feasibility of using a related, nontoxic gene element as a marker to allow monitoring and tracking of GMO products. But at a 165-nation working group that met in Chiba, Japan, on April 3, 2001, expert panels expressed concern that such a tracking policy for GMOs would put prohibitive strains on the industry.

This traceability issue is greatly magnified by the current dilemma of the new worldwide contamination with StarLink corn. The breadth of this problem was highlighted by the discovery in 2001 of widespread contamination of Mexican corn with Bt-containing substitutes. Ironically, the Mexican government had acted to prevent the potential subversion of its center of corn diversity by alien genes by banning the growing of all genetically engineered varieties of corn. Native corn plants like *chorillo* were cross-contaminated with Bt corn, during the 2000–2001 growing season, pointing to the urgency of limiting horizontal gene flow and resulting contamination. Clearly, the means to do so would involve some form of tracking.

In 1998, in our book *Against the Grain,* Britt Bailey and I argued for the need for such tracking. In 1999 and 2000, many other members of the activist community added their voices, arguing that without traceability, the world was facing an irretrievable situation where any damage associated with a GMO-derived gene fragment would be invisible.[20] Consumer groups reinforced the argument for traceability of GMO crops by noting that accountability is a universally recognized ethic.[21] The penultimate National Research Council report *Genetically Modified Pest-Protected Plants* raised the issue of tracking and allergenicity in bold print, recommending that a formal protocol be adopted for a thorough immunological review of new products before their release for consumption.

Proponents of biotechnology originally dismissed concerns about traceability and health effects as being the result of hysteria. Others have declared Cry9C to be a "non-problem." Based on a handful of tests on just seventeen of the initial group of fifty-seven persons with alleged allergic reactions to corn-flour products made with StarLink-contaminated grain, the Centers for Disease Control and Prevention (CDC) declared no one had been adversely affected. However, possible problems of the reliability of the assay and selection bias of subjects may cloud the results. Opposition groups such as Friends of the Earth have noted failings in the test design and mechanics of the CDC's follow-up studies for the Cry9C protein.[22] But even the opposition has been unable to offer a solution to the present dilemma of mass contamination with novel genetic material.

As a counterpoise to the Cry9C portion of this agricultural debacle, Aventis, the developer of StarLink, has spent over $821 million to "buy back" contaminated corn but perhaps not before the corn seed supply itself was contaminated through cross-pollenization. Regardless of Aventis's actions, public costs of this first misadventure have been considerable. As a result of contamination of feed stocks with Cry9C, the U.S. Department of Agriculture has removed 322,000 bushels of seed corn from the market at an estimated cost of $12 million.[23] This amount is only a fraction of the 430 million bushels of corn estimated by private analysts to have been contaminated by the StarLink variety.[24] The StarLink contamination has also raised international concerns that the United States is dumping its unacceptable corn products on poor countries. The most recent example occurred in Bosnia, which was offered 40,000 metric tons of corn "for animal feed." Believing that some of this corn could get onto the black market, and from there into the human food supply, Bosnia declined the donation.[25]

Legitimate Consumer Concerns?

Anti-biotechnology activists have pointed out that people want more from their food than quantity and palatability. A recent publication that evaluated the range of consumer concerns about genetically modified food reported that some people want to have organic varieties for reasons of ecological preservation, taste, or convenience.[26] Others have expressed dismay over coercive policies.

Protesters at various meetings, such as those at the recent Biodevastation/BioJustice Conference in San Diego at the end of June 2001, have picked up this latter theme and argued they are being forced to be part of a wholesale diversion of the food supply. Many believe that the perceived disregard of social institutions and arrangements constitutes neglect of American values. In keeping with the perceived lack of regard for personal autonomy, biotechnology critics can point to major corporations running roughshod over the "rights" of conventional and organic farmers. By this account, the symbolic example of this abuse of power is passivley permitting GMO-tainted pollen to contaminate traditional crops.

Among the concerns of farmers, especially those using organic methods, are that GMO-tainted pollen readily traverses field boundaries and has begun to contaminate otherwise GMO-free crops. Since organic certification requires demonstration of being free of GMOs, this problem is a daunting one. Legally, it may be akin to the issue of chemical trespass where potentially noxious or toxic pesticides drift onto crops or land without the consent of the owner. Neither has a simple solution, but each involves a fundamental tenet of liberty: that one person's freedom can be exercised only to the extent that it doesn't impinge on the freedom of others.

Dominion over Nature: Boon or Boondoggle?

Proponents of GMOs observe that no one rose up to oppose the intentional hybridization of near relatives (e.g., in triticale wheat, where wild grass and wheat genomes were combined), or the "wild crosses" that introduced genes from distant species or viruses to impart resistance within plants against rusts and blights (as was done for the Asian pear). But these cases represent extreme examples of traditional genetic manipulation. While these genetic intrusions raise questions no less extreme than those posed by radical genetic crosses in crop development, the originally introduced genes in triticale wheat or other "extreme" hybrids are neither as alien nor as potentially promiscuous as are the genes used to create many of the GMOs in commerce today. These genes include "cassettes" from the cauliflower mosaic virus, the tumor-inducing agrobacterium; from flowers such as the petunia; and from synthetic genes. The fact that earlier events such as the introduction of viral genes into the Asian pear went largely unnoticed by members of the environmental activist community only heightens the new concern over intentional movement of genes across species lines.

The trans-specific movement of genes in any form raises novel questions of the legitimate extent of human dominion over nature. The current crude methods of insertion and incorporation of the genes used in GMOs today appear to have a destabilizing effect on plant genomes, making the evolutionary survival of new lineages questionable. The combinations of genes used to splice new properties into food crops were selected for expediency, not necessarily safety. Both antibiotic-resistant genes and portions of potentially pathogenic viral genomes (e.g., the cauliflower mosaic virus) were incorporated into gene-modified food crops with little concern for secondary effects, such as shifts in the resistance patterns of microbial members of the ecosystem or in human pathogens. Certainly, using these examples to highlight the present neglect of traditional risk assessment is a legitimate position of the opponents.

But at the root of these legalistic and ethical assertions remains a dogged and, until now, irresolvable concern that somehow genetically engineered cropping methods and the crops themselves represent a fundamental departure from what went before in our subordination of Nature. The key issues of just how GMOs are impacting on agricultural and even global ecosystems cry out for review. They include review of the new mass embarkation of genetic engineers into world agriculture and the still unanswered questions concerning the extent of review necessary to meet a minimum standard of public safety and health protection. While the existing Biosafety Protocol developed by Beth Burrows and her colleagues is a necessary first step to minimizing adverse ecosystem effects, its adaptation and implementation have been spotty and incomplete.

Clearly, genetic trespass is a global question that urgently needs attention. Currently, it is being handled only on a local and individual level as if it were solely a question of property rights. The present-day near epidemic spread of genetically engineered plants points to a fundamental weakness in this formulation. The focus on individual and uniquely Western biases about the primacy of personal autonomy in litigation misses the larger geopolitical issues of sovereignty and national control over indigenous agriculture. Nothing less than global concern over the potential for genetic and evolutionary disruption within the plant kingdom is needed to prevent the recent episodes in Mexico and India from being replicated on a greater scale.

To date, no one has completed the kind of inventory needed to assess the intrinsic danger to species integrity through the loss of genetic diversity that has accompanied such episodes, or the impact on sustainable agriculture of the wholesale substitution of engineered for conventional varieties of crop plants. Nor has anyone proposed to track the spread of alien genes into the biosphere.

A Global Ethical Perspective

We who question the speed and scale of introduction of these new genetic technologies need to develop these concepts into a new world-view. We need an adventurous and solid plan leading to sustainable agricultural practices—one that, accounting for world hunger and supporting the indigenous sovereignty over agricultural practices—either excludes biotechnology or shows how it can safely be incorporated into a systematic plan. If opponents to this new technology remained fixed on the issue of personal convenience (e.g., labeling), or insist on making the issue one of corporate dominance, concerns over the introduction of GMO foods will remain largely a political issue. As such, it can be largely addressed by examining the politics and legal aspects of monopoly. But if as I believe what is really at stake is something more fundamental, affecting ownership of germ plasm and intergenerational effects, it requires a very different analysis.

A key issue is the loss of crop diversity. Such an issue requires an analysis of food crop genetics, perhaps comparing a new set of seed catalogs to their older counterparts and then developing a full, evolutionary perspective on the minimum diversity needed to sustain a cultivar line. We would note where there are fewer cultivars of certain traditional crops available today than were available ten years ago. To make this analysis complete, we would need to document why such a reduction in crop types is "bad" from an evolutionary viewpoint. Secondly, we would have to expand the potential panoply of health risks, so readily dismissed by both BIO and the NRC, to include the subtle impacts of biotechnology on the gut microflora or shifting nutritional problems. A full-blown panel review of how the subtle changes, if any, in dietary GMOs and their effects on children in particular is needed.

As an example, the new efforts of Monsanto to increase the phytoestrogens levels in soybeans through genetic engineering might prove harmful, especially if the more estrogenic ones contributed to a later adult risk of uterine cancer, a finding suggested by a new report on the results of giving genistein, the closely related aglycone of genestin, to newborn mice.[27] Monsanto appears to be on the horns of a dilemma here. By "improving" the phytoestrogen content, it may eventually have a crop with more heart-healthy properties, or at least one that can be more effectively extracted to make high-potency supplements. But Monsanto is also courting disaster should it prove true that the same phytoestrogens can be carcinogenic to infants or children at high doses. Given the scale of the operation of agricultural biotechnology and the inevitable population-wide impact of its products, simple proportionality would ask for a more thoroughgoing assessment of this kind of impact. Ideally, that assessment would have preceded the large-scale introduction of food crops across the landscape. With hindsight, ethics demands we at least provide a systematic assessment of the possible adverse effects, and benefits, of this new technology, if for no other reason than populations (both current and future) have not given their consent to this new, quasi-experimental enterprise.

A preliminary move toward examining the global impact of GMOs has recently appeared in print. In a study sponsored by the American Association for the Advancement of Science, two researchers examined thirty-five peer-reviewed studies and looked at the GMO risk of creating superweeds, new viral diseases, and unintended harm to nonpest species (e.g., the monarch butterfly).[28] They found that GMO crops held potential for both risk and benefit, but acknowledged that scientists had done little to assess the *long-term* environmental consequences of a wholesale adoption of newly engineered crops.

Clearly, it would be inappropriate to assume that people have given their passive ethical assent to a new crop simply because they are buying products that may be rife with indeterminate amounts of engineered corn or soybeans. On a more ominous note, it may be an even greater presumption to assume that the ecosystem is "safe" from Roundup Ready or other pesticide-dependent GMOs simply because only a few new weedy species have escaped, and "only" a handful of microorganisms appear affected by the persistent presence of one variety of chemical in the soil. To assert that the new weeds, such as Roundup-tolerant Australian rye grass, will have no real chance of marching through an ecosystem since their new "advantage" depends on farmers using Roundup outside their fields is to miss a salient point: Roundup Ready crops were engineered to overproduce key enzymes needed in photosynthesis. The presence of these genes in an unintended species may be a formula for a new GMO weed akin to the modern scourge produced by kudzu vines in the southern United States.

A report from the Advisory Committee charged with advising the USDA on agricultural biotechnology recently underscored several positive principles that should be followed as biotechnology expands its sphere of influence. These included the necessity to sustain and enhance crop biodiversity; to strike a balance between the public and private sectors in breeding new crops; to assure the free flow of knowledge within the research community; and to assure public confidence in new technologies.[29] None of these objectives has been subject to the public review needed to meet the requirements of public consent—namely, representative review, a bona fide risk-benefit analysis, and a consensual mechanism to assure regulatory review allows public input.

Until now, each of these principles, the activist community may properly observe, has largely been honored in the breech. First, the agricultural biotechnology industry has failed to factor in crop biodiversity in its initial takeover of major seed companies. Those we visited and researched, such as Hartz Seed in Stuttgart, Arkansas, initially and rapidly replaced a wide crop base (in the case of soybean cultivars) with a much more limited variety of engineered varieties. Second, the industry research sector, in our experience, has been loathe to share its findings with researchers in the public sector. We have been repeatedly rebuffed in our efforts to obtain any seed for testing once we shared our initial findings on soybean composition with Monsanto scientists. And third, the public confidence in GMOs is still in a fluid stage despite the massive public relations budget slated for improving the acceptance of agricultural biotechnology (estimated at some $40 million in 2000). Given the absence of labeling, lack of public accountability, and lack of free exchange of research data, biotechnology's claim to have followed "best practices" not just for business but for ethics cannot be accepted.

Ultimately, to fully assess the impact of biotechnology on both the environment and food, I would argue that we may need the social equivalent of an Environmental Impact Statement. To account for the scale at which biotechnology is currently operating, also needed would be something akin to a Societal Impact Study. Only if we examine agricultural biotechnology's impact along the axis of both its social/political acceptability and its public health benefit/risk will we gain an appreciation of its technological advances.

In addition, an ethical analysis that includes a universal index of social dependency and personal autonomy and reviews the issue of "exit" would help to establish the long-term acceptability—once relative safety were assured—of GMOs. Finally, the advent of GMOs offers a new opportunity to reexamine the issue of sustainability. While the shift to industrialized agriculture has not occurred solely as a result of the introduction of biotechnology-developed seeds, agricultural biotechnology's role in driving farming still further away from sustainable agricultural practices needs examination.

Cultural Perspectives

What we are witnessing in the "biotechnology revolution" of agriculture may well be the replacement of an old culture with a new one. The culture that was idealized in the minds of our forefathers, including "getting your hands dirty" and living close to the land, has been displaced with laser-driven sprinkler systems and satellite-assisted farming schedules. Heirloom varieties and ecologically placed cultivars of food crops have been largely supplanted with genetically controlled species, with little if any natural variation. As seen in Mexico, some seed banks may have lost some of their integrity through the incursion of engineered genetic varieties. And the full range of crop biodiversity may be greatly reduced by substitution with a limited number of engineered varieties.

Ultimately, we have little idea how the present GMO-dominated culture will impact wildlife, natural diversity, or our own well-being— just as we had no idea how the previous reliance on chemical farming would decimate species, contaminate water, and create dependency on commercial-scale farming methods. In the trend for wholesale corporate dominance to substitute for more sustainable farming, we may be witnessing a further decay in ecosystem stability. Some opponents, like many of the contributors in this volume, have spoken critically about the loss of local control, the end of the small farmer, and the privatization of the genetic commons through outright ownership of living forms and the patenting of life. But the larger evolutionary issues deserve a forum of their own.

Biotechnology companies may have taken a step too far in putting agriculture into a monopolistic mode without this global perspective. In one sense, the consequences of agricultural biotechnology may be no different than those produced by other instances of corporate power

that strive for dominance without accountability. If we are indeed seeing the emergence of a new genetic-industrial complex, we should be
concerned about the scale and impact of this coordinated effort.
Biotechnology is merely an extreme expression of an existing social and
political trend that can be harnessed for good or ill. Will this new culture create novel forces of dependency? Will the acceptance of GMO
foods become a *fait accompli*? And if so, will that acceptance simply
reinforce an existing trend toward engineering and mass production on
the farm?

As Richard Lewontin points out, since the early days of mechaniza-
tion, farmers have never had a meaningful "exit" from the dominance
of technology. In his words, "For the farmer there is no escape from
engineering, whether it be mechanical, electrical, or genetic."[30] This
view adopts too cynical an interpretation of technology. Industrialization of agriculture in one form and not another is no more inevitable
than is the creation and expansion of organic farming. Each requires
public acquiescence and support. Each requires a battle for the public
mind and heart over the acceptability of methods of production,
genetic manipulation, and mass production of food products.

Efficiency and scale alone do not determine success. The present
trend toward mass expansion of certain crops such as those containing
Bt creates a "political reality on the ground." If we continue to allow so
many crops to become genetically engineered that cross-pollination
becomes a fact of life, an inadvertent consequence (or a plan) of the
success of agricultural biotechnology companies may lead to the dominance of GMOs—and, in the extreme case, create ecologic havoc.
Regardless of the origins of the present ubiquity of GMOs, this trend
toward globalization of promiscuous GMO crops carries grave moral
implications—not just for the demise of the small farmer, but for the
substitution of a new political maxim without public consent that
threatens agricultural stability, local control, and potentially the
integrity of centers of crop biodiversity themselves.

Political power no longer comes through the barrel of a gun—it
comes from control of food crops and their seeds, and ultimately the
indigenous genetics of crops chosen by the people. Right now it is
plants that are under intense geopolitical pressure. In the future it may
be people themselves. Unless we get the rules straight now for how such
megalithic control can be tempered and reconciled with human values,
we may be in for a long political struggle indeed.

Afterword: The Biotech Distraction

Frances Moore Lappé

Biotechnology companies and even some scientists argue that we need genetically modified seeds to feed the world and to protect the earth from chemicals. Their arguments feel eerily familiar.

Thirty years ago, I wrote *Diet for a Small Planet* for one reason. As a twenty-six-year-old self-taught researcher buried in the U.C. Berkeley agricultural library, I was stunned to learn that the experts—equivalent to the biotech proponents of today—were wrong. They were telling us we'd reached the earth's limits to feed ourselves, but in fact there was more than enough food for us all.

Hunger, I learned, is the result of economic "givens" we ourselves have created, assumptions and structures that actively generate scarcity from plenty. Today this is more, not less, true.

Throughout history, ruminants had served humans by turning grasses and other "inedibles" into high-grade protein. They were our four-legged protein factories. But once we began feeding livestock from cropland that could grow edible food, we began to convert ruminants into our protein disposals. Only a small fraction of the nutrients fed to animals return to us in meat; the rest, animals use largely for energy or they excrete. Thirty years ago, one-third of the world's grain was going to livestock; today, it is closer to one-half. And now we're mastering the same disappearing trick with the world's fish supply. Feeding fish to fish, again we're reducing potential supply.

We're shrinking the world's food supply for one reason: The hundreds of millions of people who go hungry cannot make sufficient "market demand" for the fruits of the earth. So more and more of it flows into the mouths of livestock, which convert it into what the better-off can afford. Corn becomes filet mignon. Sardines become salmon.

Enter biotechnology. While its supporters claim that seed biotechnology methods are "safe" and "precise," other scientists strongly refute that, as they do claims that biotech crops have actually reduced pesticide use. But this very debate is in some ways part of the problem. It is a tragic distraction our planet cannot afford.

We're still asking the wrong question! Not only is there already enough food in the world, but as long as we are only talking about food—how best to produce it—we'll never end hunger or create the communities and food safety we want.

We must ask instead: How do we build communities in tune with nature's wisdom in which no one, anywhere, has to worry about putting food—safe, healthy food—on the table? Asking this question takes us far beyond food. It takes us to the heart of democracy itself, to whose voices are heard in matters of land, seeds, credit, employment, trade, and food safety.

The problem is, *this* question cannot be addressed by scientists or by any private entity, including even the most high-minded corporation. Only citizens can answer it, through public debate and the resulting accountable polities that come from our engagement.

Where are the channels for public discussion and where are the accountable polities? Increasingly, public discussion about food and hunger is framed by advertising from multinational corporations that increasingly control not only food processing and distribution but farm inputs and seed patents. Two years ago, the seven leading biotech companies, including Monsanto, teamed up under the neutral-sounding Council for Biotechnology Information and are spending millions to, for example, blanket us with full-page newspaper ads about biotech's virtues.

Increasingly, polities are more beholden to these corporations than to their citizens. Nowhere is this more dramatically true than in decisions regarding biotechnology—whether it's in the government's approval and even patenting of biotech seeds and foods without public input, or in the rejection of mandatory labeling of biotech foods despite broad public demand for it.

The absence of genuine democratic dialogue and accountable government is a prime reason most people remain blind to the many breakthroughs in the last thirty years demonstrating we *can* grow abundant, healthy food and protect the earth. We can successfully work with

Nature rather than crossing boundaries Nature would never cross, as with biotechnology.

Hunger is not caused by a scarcity of food but by a scarcity of democracy. Thus it can never be solved by new technologies, even if they were to be proven "safe." It can only be solved as citizens build democracies in which government is accountable to them, not private corporate entities.

Acknowledgment

This Afterword appears with permission from Frances Moore Lappé. It originally appeared in the June 27, 2001, edition of the *Los Angeles Times*.

In Defense of the
Precautionary Principle

Carolyn Raffensperger and Katherine Barrett

> When an activity raises threats of harm to the environment or human health, precautionary measures should be taken even if some cause and effect relationships are not fully established scientifically.
>
> —Wingspread Statement on the Precautionary
> Principle, January 1998

The precautionary principle was first established as a concept of environmental law in the 1970s. Since that time, precaution has been invoked in numerous international environmental agreements, including the 1992 Rio Declaration on Environment and Development, and more recently the Cartagena Protocol on Biosafety, which regulates international movement of genetically modified (GM) organisms. The precautionary principle is also stated explicitly in the environmental policies of several countries (e.g., Canada, Australia, and Sweden) and in the Maastricht Treaty of the European Union. The U.S. Department of Agriculture (Washington, DC) and the Food and Drug Administration (FDA, Rockville, MD) adamantly claim that U.S. food safety policies are firmly grounded in a precautionary approach, but stop short of acknowledging precaution as a principle of law.[1] Precaution, therefore, is a widely recognized and adopted foundation for making wise decisions under uncertain conditions.

Although there are differences in wording, three core elements are present in all statements of the precautionary principle: If there is reason to believe that a technology or activity may result in harm, and there is scientific uncertainty regarding the nature and extent of that harm, then measures to anticipate and prevent harm are necessary and justifiable.

The precautionary principle is necessary and justifiable because, simply stated, our ability to predict, calculate, and control the impacts of technologies such as GM organisms is limited. The novelty and complexity associated with inserting isolated gene constructs into organisms, and releasing those organisms on a global scale, demand that we acknowledge uncertainties, accept responsibility, and exercise due caution.

This is recognized by the international adoption of the Protocol on Biosafety and by independent scientific bodies in the United States, the European Union, and Canada, among others.[2,3]

Although there is consistency among definitions, no uniform, global recipe exists for implementing the precautionary principle. It is a general principle, not a set of rules, and it must remain responsive to social and ecological context. Nonetheless, it is possible and important to set procedural guidelines such that implementation is not arbitrary. We advocate the following six steps:

- Set broad social, environmental, and economic policies that outline clear, long-term goals. For example, how can we achieve environmentally, economically, and socially sustainable agriculture?
- Assess alternatives. Are GM crops necessary for reaching defined goals? Are there more beneficial, less uncertain, and less controversial ways to achieve individual and collective goals?
- Define parameters of "potential harm" for all potential alternatives, including long-term, cumulative, synergistic, and indirect harms to both ecological and social systems.
- Analyze the sources and extent of uncertainty, including gaps in scientific data, inadequate methods to predict impacts, the intractability of confluent complex systems, and uncertainties created through insufficient funding for risk-related studies.
- Weigh evidence from diverse sources, including peer-reviewed scientific research and the experience-based knowledge of people directly involved in the issues.
- Adopt appropriate precautionary actions, which may range from a complete ban or phase-out, to moratoria, to conditional approvals with provisions for monitoring and feedback.

By this process, the precautionary principle is neither unscientific nor anti-technology. It requires robust scientific analysis with close attention to uncertainty and to the probability of both false positive and false negative conclusions. The precautionary principle can also stimulate alternative directions for regulatory policies and technology development. Its power lies not in halting all new activities, but in heightening our attention to the potential consequences of our actions, shifting the scope of questions we ask about technologies, and finding innovative solutions to complex problems.

Above all, the precautionary principle is grounded firmly in democratic process. None of the above steps can be implemented without transparent and inclusive decision making.

Lack of democratic process has been a primary source of contention surrounding GM crops and food. Under the precautionary principle, not only is this ethically unacceptable, it is an impoverished procedure for making decisions about a technology that now affects (voluntarily or not) millions of people and many other species throughout the world.

Acknowledgments

This letter appears by permission of the authors and *Nature Biotechnology*, September 2001, Vol. 19, No. 9, pages 811–812.

Carolyn Raffensperger is executive director of the Science and Environmental Health Network www.sehn.org. Katherine Barrett is project director of the Science and Environmental Health Network, University of Victoria, Canada.

Appendix B

A Declaration of Bioethics and Agricultural Biotechnology

Marc Lappé and Phil Bereano

The following declaration was written by Marc Lappé and Phil Bereano, Professor of Technology and Communication at the University of Washington, with input from participants following a conference of environmentalists, ethicists, and philosophers in Bolinas, California, in June 1999. The declaration embodies many of the philosophical and ethical principles enumerated in this book. It puts into capsular form many of the concerns about who controls the genetics of food crops and what the burden of proof should be for establishing the safety of newly engineered ones.

The key elements of the declaration include the requirement for thorough testing; a "general welfare provision" that asserts engineered crops serve the common good; an assertion that the locus of control should remain with farmers; and that full disclosure of risks (and benefits) of crops and their health and nutritional properties, where known, be widely available.

The declaration also reasserts the provision that the consumer must be given the right to refuse participation in the new system of genetic engineering; that labeling is a consumer right that can be exercised; and that the precautionary principle of not harming be given primacy. The declaration also decries the current lack of testing, because it limits the consumer's confidence in food safety and puts at risk present and future generations of persons who may be especially vulnerable to adverse effects from any new proteins of gene-directed changes in the

:sulting food. Finally, the declaration states the primary good of having the atents for engineered crops remain in the public domain.

The ten points of the declaration encapsulate many of the positions and :atements that appear throughout the present book. It was signed by more than ighty groups, as shown in the accompanying list of organizations that endorse 1e Pacific Declaration.

The Pacific Declaration

We the undersigned, in recognition of the fundamental importance of our planet's 1tural genetic heritage and diversity, and in acknowledgment of the power of genetic 1gineering to transform this heritage, believe that the proponents and practitioners of :netic technologies must adhere to the principles of prudence, transparency and :countability.

We also aver that respect for life, ensuring a habitable planet, and protecting ecosys-ms are universally recognized and fundamental human values. For this reason, those tering the genetic integrity of natural species bear the burden of proving their inter-ntions will not jeopardize these values.

We also believe in democracy. In democratic societies, any decision to deploy power-l new technologies must be made with full public participation and accountability. » date, our government, international agencies, public universities, and biotechnology rporations have neglected these objectives. Therefore we declare:

. Environmental safety and public health require the systematic study of any trans-genically modified living organism over multiple generations before allowing its envi-ronmental release or marketing;

. All proposed products derived from genetic engineering must be shown to con-tribute to the general welfare of consumers, farmers, and society without compro-mising the viability of traditional agricultural practices, including organic farming;

. Farmers and agrarian peoples generally who have cultivated, nurtured, and devel-oped crops have the right to control their crop materials;

. Such control includes the right to cultivate indigenous or conventional species using traditional methods, and freely to use or re-use any genetic seed stock;

. People should have access to all relevant data concerning the potential effect of genetically modified organisms on the health of present or future generations;

. People have the right to accept or decline any food product for personal, religious, or philosophical reasons;

. In the absence of compelling evidence showing the equivalence and safety of genet-ically engineered compared to conventional foods, all food products derived from genetic technologies must be accurately labeled;

. The medical injunction to "do no harm" requires adequate and sufficient pre- and post-market testing and surveillance of genetically engineered products; The present lack of such testing contravenes this injunction and thereby jeopardizes universal access to safe food, potentially putting at risk present and future vulnera-ble populations including pregnant women and young children; and

. Because the fundamental discoveries of genetic engineering were developed through public funding, justice requires that any and all risks, costs, and benefits of the prod-ucts of genetic engineering be equitably distributed in society.

Until we have guarantees and assurances that the above stated requirements and objectives are no longer compromised by government and industry practices; and

Until our government has created a comprehensive and effective regulatory system for all products of genetic engineering; and

Until such fundamental and constitutionally guaranteed protections of life and liberty, as well as protection of the health of the environment, food security, and consumer right to know are vouchsafed;

We call upon our governmental representatives to suspend any further introduction of genetically engineered organisms and to hold the practitioners of genetic engineering, whether they be corporations, universities, or governmental agencies, fully liable for any adverse consequences of their work.

Notes

Introduction. GMOs, Luddites, and Concerned Citizens

1. T. Gura, "The battlefields of Britain," *Nature* 412 (2001): 760–763.
2. R. Dalton, "Transgenic corn found growing in Mexico," *Nature* 413 (2001): 337.
3. European Commission. *Official Journal* L006 (2000): 13–14.
4. S. Kay and G. Van den Eedge, "The limits of GMO detection," *Nature Biotechnology* 19 (2001): 405.
5. See http://news.excite.com/news/pr/010619/ny-harris-interactive.html.

Chapter One. Ethical Issues Involving the Production, Planting, and Distribution of Genetically Modified Crops

1. S. Krimsky, *Genetic Alchemy* (Cambridge, MA: The MIT Press, 1982).
2. J.E. Carpenter and L.P. Gianessi, *Agricultural Biotechnology: Updated Benefit Estimates* (Washington, DC: National Center for Food and Agricultural Policy, 2001), 1.
3. M. Midgley, "Biotechnology and monstrosity," Hastings Center Report 30 (2000): 7–15.
4. Committee on the Introduction of Genetically Engineered Organisms into the Environment, National Academy of Sciences, *Introduction of Recombinant DNA-Engineered Organisms into the Environment: Key Issues* (Washington, DC: National Academy Press, 1993).
5. Committee on Scientific Evaluation of the Introduction of Genetically Modified Microorganisms and Plants into the Environment, National Academy of Sciences, *Field Testing Genetically Modified Organisms: Framework for Decisions* (Washington, DC: National Academy Press, 1989).
6. Committee on Genetically Modified Pest-Protected Plants, Board on Agriculture and Natural Resources, National Research Council, National Academy of Sciences, *Genetically-Modified Pest-Protected Plants: Science and Regulation* (Washington, DC: National Academy Press, 2000).

7. Committee on Scientific Evaluation of the Introduction of Genetically Modified Microorganisms and Plants into the Environment, National Academy of Sciences, *Field Testing Genetically Modified Organisms: Framework for Decisions* (Washington, DC: National Academy Press, 1989), 64.

8. Committee on Scientific Evaluation of the Introduction of Genetically Modified Microorganisms and Plants into the Environment, National Academy of Sciences. *Field Testing Genetically Modified Organisms: Framework for Decisions* (Washington, DC: National Academy Press, 1989).

9. J.A. Nordless, S.L. Taylor, J.A. Townsend, L.A. Thomas, and R.K. Bush, "Identification of a Brazil-nut allergen in transgenic soybeans," *New England Journal of Medicine* 334 (1996): 688–692.

10. K. Eichenwald, G. Kolata, and M. Petersen, "Biotechnology food: From the lab to a debacle," *New York Times* 25 January (2001): A1, C6.

11. T.J. Hoban and P.A. Kendall, "Consumer attitudes about food biotechnology," *Project Report* (Department of Food Science and Human Nutrition, Colorado State University, 1993).

12. S. Martin and J. Tait, "Release of genetically modified organisms: Public attitudes and understanding," *Report to the Laboratory of the Government Chemist: Bioscience and Innovation* (Center for Technology Strategy. Milton Keynes, U.K., June 1992).

13. J. Walsh, "Brave new farm," *Time* 11 January (1999): 87. The article reported a public opinion survey in which the response to the question "Should genetically engineered foods be labeled as such?" was 81 percent affirmative, 14 percent negative.

14. U.S. Department of Health and Human Services, Food and Drug Administration, "Guidance for industry. Voluntary labeling indicating whether foods have or have not been developed using bioengineering," *Federal Register* 65 (2000): 56458.

15. U.S. Department of Health and Human Services, Food and Drug Administration, "Food labeling: Foods derived from new plant varieties," *Federal Register* 58 (1993): 25838.

16. U.S. FDA, 2000.

17. U.S. FDA, 2000.

18. J. Paulson, "NFU fights 'genetic pollution': National farm group wants Ottawa to make ag-biotech firms liable," *The Saskatoon Star Phoenix* 12 May (1999): 5.

19. K.S. Betts, "Growing evidence of widespread GMO contamination," *Environmental Science and Technology* 33 (1999): 484A–485A.

20. A.F. Deshayes, "Environmental and social impacts of GMOs: What have we learned from the past few years," *The Biosafety Results of Field Tests of Genetically Modified Plants and Microorganisms* (Proceedings of the 3rd International Sym-

posium, Monterey, California, November 13–16, 1994. D.D. Jones, ed. Oakland, CA: Division of Agriculture and Natural Resources, University of California, 1994), 5–19.

21. A.M. Timmons, Y.M. Charters, J.W. Crawford et al., "Risks from transgenic crops," *Nature* 380 (1996): 487.

22. U.S. Department of Agriculture, "National Organic Program. Final Rule," *Federal Register* 65 (2000): 80597–80646. http://www.ams.usda.gov/nop.

23. S.B. Levy, *The Antibiotic Paradox* (New York: Plenum, 1992).

24. M.J. Nash, "Grains of hope," *Time* 31 July (2000): 39–46.

25. V. Azais-Braesco and G. Pascal, "Vitamin A in pregnancy: Requirements and safety limits," *American Journal of Clinical Nutrition* 71 (2000): 1325S–1333S.

26. R. Paarlberg, "Agrobiotechnology choices in the developing countries," *International Journal of Biotechnology* 2 (2000): 164–173.

27. A. Coghlan, "Judging gene foods: An impartial panel could quell health and environmental fears," *New Scientist* 166 (2000): 4.

28. J. Fagan, "Science-based precautionary approach to the labeling of genetically engineered foods," http://www.naturallaw.org.nz/genetics/papers/F_label.asp.

29. K. Eichenwald, G. Kolata, and M. Petersen, "Biotechnology food: From the lab to a debacle," *New York Times* 25 January (2001): A1, C6.

Chapter Two. Why Food Biotechnology Needs an Opt Out

1. I have chosen to write this chapter in the style of popular philosophy, rather than that of a scientific article. Rather than disrupt the flow of the argument in this chapter, I have included only a few references to scholarly literature. I refer readers to two peer reviewed publications, Thompson (1997) and Thompson (2000) that are extensively referenced. These two publications contain numerous citations that document studies of public opinion on biotechnology, as well as policy analysis and philosophical studies of issues discussed throughout this chapter.

2. This is the language of the Nuffield Council on Bioethics (1999), for example, and also of a U.S. Congressional Research Service document that quotes my own work (see Vogt and Parrish 1999).

3. The term "rights" came into frequent use during the seventeenth century, when philosophers such as Hugo Grotius (1583–1645) and John Locke (1632–1704) argued that all men were created with the capacity for free and rational thought (the gender-specific term here is used advisedly). They believed that the rights needed to exercise this capacity were the moral foundation for social order. "Rights philosophy" was particularly important in the twentieth century, as documents such as the International Declaration of Human Rights became the basis for attempts to end repressive governments around the world

(see Donnely 1989). Major twentieth-century philosophical works on rights include Hohfield (1923), Dworkin (1977), and Thomson (1990). McGinn (1994) argues against the use of rights to interpret complex technological risks, while Westra (2001) argues strongly in favor of their use.

4. The magazine *Farming,* published deep in the heart of Ohio's Amish country, carries a note saying that it will not accept advertising for products containing alcohol, tobacco, or margarine.

5. Some of the key literature on this point is too recent to have been cited in the sources referenced in note 1. See, in particular, Jackson (2000), Rippe (2000), and Fraser (2001).

Bibliography

V. Fraser, "What's the moral of the GM food story?," *Journal of Agricultural and Environmental Ethics* 14 (2001): 147–159.

Nuffield Council on Bioethics, *Genetically Modified Crops: The Ethical and Social Issues* (London: Nuffield Council on Bioethics, 1999).

D. Jackson, "Labeling products of biotechnology: Towards communication and consent," *Journal of Agricultural and Environmental Ethics* 12 (2000): 319–330.

K.P. Rippe, "Novel foods and consumer rights: Concerning food policy in a liberal state," *Journal of Agricultural and Environmental Ethics* 12 (2000): 71–80.

P.B. Thompson, *Food Biotechnology in Ethical Perspective* (London: Chapman and Hall [now distributed by Aspen Publishing], 1997).

P.B. Thompson, *Food and Agricultural Biotechnology: Incorporating Ethical Consideration* (Ottawa: Canadian Biotechnology Advisory Committee, 2000), 40. Also available in French: *Intégration de facteurs d'éthique à la biotechnologie alimentaire et agricole* (Ottawa: Comité consultatif Canadien de la biotechnologie, 2000). English text available online at: http://www.agriculture.purdue.edu/agbiotech/Thompsonpaper/Canadathompson.html

D.U. Vogt and M. Parrish, *Food Biotechnology in the United States: Science, Regulation, and Issues* (Washington, DC: Congressional Research Service, 1999).

J. Donnelly, *Universal Human Rights in Theory and Practice* (Ithaca, NY: Cornell University Press, 1989).

R.M. Dworkin, *Taking Rights Seriously* (Cambridge, MA: Harvard University Press, 1977).

W.N. Hohfield, *Fundamental Legal Conceptions* (New Haven, CT: Yale University Press, 1923).

R. McGinn, "Technology, demography and the anachronism of traditional rights," *Journal of Applied Philosophy* 11.1 (1994).

J.J. Thomson and J. Jarvis, *The Realm of Rights* (Cambridge, MA: Harvard University Press, 1990).

L. Westra, "Environmental risks, rights and the failure of liberal democracy," in L. Pojman, ed., *Environmental Ethics: Readings in Theory and Application*, 3rd ed. (Belmont, CA: Wadsworth Publishing, 2001), 515–527.

Chapter Three. A Naturalist Looks at Agricultural Biotechnology

1. B. Johnson, "The ethical considerations of genetic modification," *SPLICE* (1999).

2. W. Cronon, *Changes in the Land: Indians, Colonists, and the Ecology of New England* (New York: Hill and Wang, 1983).

3. H. Brody, *The Other Side of Eden* (Vancouver, BC: Douglas & McIntyre, 2000).

4. B. Kingsolver, *Prodigal Summer* (New York: HarperCollins, 2000).

5. M. Talbot, "A desire to duplicate," *New York Times Magazine* (2001): 67.

6. D. Cayley, CBC Radio Series IDEAS, *The Corruption of Christianity* (with Ivan Illich), CBC (2000): 4, 6, 38.

7. R. Lewontin, *The Triple Helix* (Cambridge, MA: Harvard University Press, 2000), 54.

8. B. Knoppers, "Considering our genetic legacy," *Nature Biotechnology* 18 (2000): 1115.

9. M. Somerville, *The Ethical Canary* (Toronto: Viking Press, 2000), x.

10. F. Koechlin, Blue Ridge Institute, Switzerland, *Mail-out*, Dec. 2000 / January 2001 (E-mail).

11. An interesting apparent reversal in this process of shifting the burden of proof appeared the "Regulations Amending the Seeds Regulations" of the "Seeds Act and Canada Agricultural Products Act." The changes were discussed in a Regulatory Impact Analysis Statement published in the *Canada Gazette* (#44, 28/10/00). "The proposed regulatory amendments will clearly put the onus on those persons being authorized to conduct confined field trials or persons carrying out the field trials on their behalf . . . to prevent PNTs from becoming livestock feed or entering any food for humans." This apparent change of policy is notable for its proximity to the Aventis StarLink fiasco. Apparently the regulatory agency—the Canadian Food Inspection Agency, in this instance—came to the realization that it did not want to be held responsible for the acts of the biotech industry, having observed what was happening in the United States.

Chapter Four. When Transgenes Wander, Should We Worry?

1. L. Hall, K. Topinka, J. Huffman, L. Davis, and A. Allen, "Pollen flow between herbicide-resistant *Brassica napus* is the cause of multiple-resistant *B. napus* volunteers," *Weed Science* 48 (2000): 688–694.

2. J. Gressel, "Tandem constructs: Preventing the rise of superweeds," *Trends in Biotechnology* 17 (1999): 361–366.

3. N.C. Ellstrand, "Gene rustlers," *Omni* 11.7 (1999): 33.

4. A.A. Snow and P. Palma, "Commercialization of transgenic plants: Potential ecological risks," *BioScience* 47 (1997): 86–96. See also R.S. Hails, "Genetically modified plants: The debate continues," *Trends in Ecological Evolusion* 15.222 (2000): 14–18.

5. J. Rissler and M. Mellon, *The Ecological Risks of Engineered Crops* (Cambridge, MA: The MIT Press, 1996).

6. Scientists' Working Group on Biosafety, *Manual for Assessing Ecological and Human Health Effects of Genetically Engineered Organisms, Part One: Introductory Materials and Supporting Text for Flowcharts, and Part Two: Flowcharts and Worksheets.* (Edmonds, WA: The Edmonds Institute, 1998).

7. R.E. Colwell, E.A. Norse, D. Pimentel, F.E. Sharples, and D. Simberloff, "Genetic engineering in agriculture," *Science* 229 (1985): 111–112.

8. N.C. Ellstrand, "Pollen as a vehicle for the escape of engineered genes? "In J. Hodgson and A.M. Sugden, eds. *Planned Release of Genetically Engineered Organisms* (Cambridge, UK: Elsevier, 1988), S30–S32.

9 P.J. Dale, "Spread of engineered genes to wild relatives," *Plant Physiology* 100 (1992): 13–15.

10. R.M. Goodman and N. Newell, "Genetic engineering of plants for herbicide resistance: Status and prospects, " In H.O. Halvorson, D. Pramer, and M. Rogul, eds. *Engineered Organisms in the Environment: Scientific Issues* (Washington, DC: American Society for Microbiology, 1985), 47–53.

11. T. Klinger, D.R. Elam, and N.C. Ellstrand, "Radish as a model system for the study of engineered gene escape rates via crop-weed mating," *Conservation Biology* 5 (1991): 531–535.

12. T. Klinger and N.C. Ellstrand, "Engineered genes in wild populations: Fitness of weed-crop hybrids of radish, *Raphanus sativus," L. Ecological Applications* 4 (1994): 117–120.

13. P.E. Arriola and N.C. Ellstrand, "Crop-to-weed gene flow in the genus *Sorghum* (Poaceae): Spontaneous interspecific hybridization between johnsongrass, *Sorghum halepense,* and crop sorghum, *S. bicolor," American Journal of Botany* 83 (1996): 1153–1160.

14. P.E. Arriola and N.C. Ellstrand, "Fitness of interspecific hybrids in the genus *Sorghum:* Persistence of crop genes in wild populations," *Ecological Applications* 7 (1997): 512–518.

15. N.C. Ellstrand, H.G. Prentice HC, and J.F. Hancock, "Gene flow and introgression from domesticated plants into their wild relatives," *Annual Review of Ecology and Systematics* 30 (1999): 539–563.

16. Ibid.

17. Ibid.

18. Ibid.

19. I.M. Parker and D. Bartsch, "Recent advances in ecological biosafety research on the risks of transgenic plants: A transcontinental perspective," In J. Tomiuk, K. Wohrmann, and A. Sentker, eds. *Transgenic Organisms: Biological and Social Implications* (Basel: Birkhauser Verlag, 1996), 147–161.

20. N.C. Ellstrand and D.R. Elam, "Population genetic consequences of small population size: Implications for plant conservation," *Annual Review of Ecology and Systematics* 24 (1993): 217–242.

21. G.R. Huxel, "Rapid displacement of native species by invasive species: Effect of hybridization," *Biological Conservation* 89 (1999): 143–152.

22. D.E. Wolf, N. Takebayashi, and L.H. Rieseberg, "Predicting the risk of extinction through hybridization," *Conservation Biology* 15 (2001): 1039–1053.

23. E. Small, "Hybridization in the domesticated-weed-wild complex," In W.F. Grant, ed, *Plant Biosystematics* (Toronto: Academic Press, 1984), 195–210.

24. Y.T. Kiang, J. Antonovics, and L. Wu, "The extinction of wild rice (*Oryza perennis formosana*) in Taiwan," *Journal Asian Ecology* 1 (1979): 1–9.

25. M. MacArthur, "Triple-resistant canola weeds found in Alberta." The Western Producer, http://www.producer.com/articles/20000210/news/20000210news01.html (February 10, 2000).

26. P. Callahan, "Genetically altered protein is found in still more corn," *Wall Street Journal* 236 (2000): B5.

Chapter Five. Patents, Plants, and People: The Need for a New Ethical Paradigm

1. This clever phrase was coined in the editorial, "The Big Test," *The New Republic*, July 10 & 17, 2000, at 9.

2. Promotions that mislead parents into thinking their children can survive on soy milk alone are a case in point.

3. L.B. Andrews, *The Clone Age: Adventures in the New World of Reproductive Technology* (New York: Henry Holt, 2000), 146.

4. S. Newman, "Don't try to engineer human embryos," *St. Louis Post-Dispatch* 25 July 2000, B15.

5. R. Shield, "A genetically enhanced harvest," *Science Spectra* 25 (2000): 18–27.

6. Monsanto Financial Summary 1998 Annual Report, Investor Relations, available at http://www.monsanto.com.

7. See Jubilee 2000/USA, available at http://www.j2000usa.org/debt/chart2.html,

citing figures from The World Bank, World Development Indicators, 1999, Washington, D.C.

8. Telephone conversation with K. Marshall, Public Affairs, Monsanto.

9. K. Bosselmann, "Plants and politics: The international legal regime concerning biotechnology and biodiversity," *Colorado Journal of International Environmental Law* 7 (1996): 111.

10. "Genotypes: Earmarked for extinction?" available at http://www.gene.ch/gentech/2000/Jul/msg00066.html. Site visited 2/2/01.

11. A. Rhodes, "Saving older seed varieties may avert global disaster," *Sustainable Farming, (Winter 1995):* available at http://www.eap.mcgill.ca/MagRack/SF/Winter%2095%20N.html. Site visited 2/2/01.

12. "Genotypes: Earmarked for extinction?", supra note 10.

13. Use of hybrid seeds to prevent saving seeds for replanting by farmers is similar to intellectual property protection efforts aimed at requiring farmers to repurchase seeds each year.

14. S. Connor, "Patent plan for breasts set to stir passions," *The Independent* (London), 19 February 1992: 3.

15. See L. Andrews, "Weird science," *Chicago Magazine* (August, 2000): 22–24.

16. J.F. Merz, M.K. Cho, M.J. Robertson, and D.G.B. Leonard, "Disease gene patenting is a bad innovation," *Molecular Diagnosis* 2 (1997): 301.

17. Ibid., 299.

18. *Davis v. Davis*, 842 S.W.2d 588, 594 (Tenn. 1992) (Lejeune refers to embryos as "tiny human beings" with a right to be born).

19. L. Andrews and D. Nelkin, *Body Bazaar: The Market for Human Tissue in the Biotechnology Age* (New York: Crown Publishing Group, 2001), 43. See also K. Painter, "More tests are finding genetic time bombs," *USA Today*, 15 August 1997: 2A.

20. D. Burk, "Lex Genetica: Governing through biological code," *Bioethics Examiner* (Winter 2001): 4.

21. R. Cook-Deegan, *The Gene Wars: Science, Politics, and the Human Genome* (New York: W.W. Norton & Company, 1994), 307.

22. *Diamond v. Chakrabarty*, 448 U.S. 303 (1980).

23. Ibid. at 309–310.

24. Ibid. at 309, citing S. Rep. No. 1979, 82d Cong., 2d Sess., 5 (1952); H.R. Rep. No. 1923, 82nd Cong., 2d Sess., 6 (1952).

25. 35 U.S.C. § 102.

26. 35 U.S.C. § 103.

27. 35 U.S.C. § 101. It should be noted, though, that the decision was made by a divided court. Five justices favored the patent. Four would have denied it. Public scrutiny of the decision was minimal. No one really cared about bacteria. Nor

did anyone at the time question the scientific hype. (The bacteria apparently does not work very well [R.L. Hotz, "Microbes sought to 'eat' waste," *Atlanta Journal-Constitution*, 17 November 1987: A22] and has never been used. R.M. Atlas and C.E. Cerniglia, "Bioremediation of petroleum pollutants," *Bio-Science* 322 [1995]: 45.) People generally did not anticipate how the decision might be applied to animals (as it was, later, to Harvard/DuPont OncoMouse) or to parts of people (like a breast cancer gene).

28. Brief on Behalf of Peoples Business Commission, *Diamond v. Chakrabarty*, No. 79-136 (December 13, 1979).

29. 35 U.S.C. § 101.

30. *Funk Bros. Seed Co. v. Kalo Inoculant Co.*, 333 U.S. 127 (1948). The decision said, "patents cannot issue for the discovery of the phenomena of nature. The qualities of these bacteria, like the heat of the sun, electricity, or the qualities of metals, are part of the storehouse of knowledge of all men. They are manifestations of laws of nature, free to all men and reserved exclusively to none." Id. at 130 (citation omitted). However, if an inventor isolates and purifies a substance so that it could be used in a way the impure product could not, that inventor can receive a patent. *In re Bergstrom*, 427 F.2d 1394 (C.C.P.A. 1970).

31. L. Roberts, "Who owns the human genome?" *Science* 358 (1987): 237.

32. Ibid.

33. P. Gorner, "Parents suing over patenting of genetic test; they say the researchers they assisted are trying to profit from a test for a rare disease," *Chicago Tribune*, 11 November 2000: 1.

34. A. Pollack, "Bristol-Myers and Athersys make deal on gene patents," *New York Times*, 8 January 2001: C2.

35. S. Mayer, B. Ayliffe, and B. Price, "Patenting life," *The Guardian* (London), 15 November 2000: 6.

36. R. Kotulak, "Taking license with your genes: Biotech firms say they need protection," *Chicago Tribune*, 12 September 1999: 1, citing a survey of 120 labs by University of Pennsylvania bioethicist J. Merz.

37. Ibid.

38. Rabbi J. Ekstein, who founded a testing program in New York, says, "If Canavan testing won't be available—which is how it looks if they enforce the patent—there's no question Canavan children will be born." J. Peres, "Genetic tests reduce neighborhood's grief," *Chicago Tribune*, 12 September 1999: 16.

39. L. Andrews and D. Nelkin, supra note 19 at 55.

40. M.A. Heller and R.S. Eisenberg, "Can patents deter innovation? The anticommons in biomedical research," *Science* 280 (1998): 698.

41. Ibid.

42. Ibid., 699.

43. The example was given by Wallace Judd, president of the California-based Mentrix Corporation. It is discussed in Seth Shulman, *Owning the Future* 67 (Boston: Houghton Mifflin, 1999).

44. R.M. Kunstadt, F.S. Kieff, and R.G. Krammer, "Are sports moves the next in IP laws?" *National Law Journal*, May 20, 1996. See also "Yo! He owns that move," *Sports Illustrated*, May 27, 1996, which provides examples of potential patents such as: "Bob Cousy BEHIND THE BACK PASS. The inventor, a fully accredited basketball professional, seeks to protect a form of offensive chicanery in which he manually maneuvers a basketball behind his back, sometimes accompanying this maneuver with deceptive movements of his head and/or eyes, and rapidly discharges said ball so as to reassign possession of it to a team-mate."

45. L. Andrews and D. Nelkin, supra note 19 at 28.

46. *Moore v. Regents of the Univ. of California* 51 Cal. 3d 120, 175 (1990) (Mosk, J., dissenting).

47 Ibid.

48. L. Andrews, "My body, my property," *Hastings Center Report* (October 1986): 28.

49. N. Roht-Arriaza, "Of seeds and shamans: The appropriation of the scientific and technical knowledge of indigenous and local communities," *Michigan Journal of International Law* 17 (1996): 919, 924.

50. Ibid.

51. C. Cray, "Neem Tree Freed," *Multinational Monitor*, 1 June 2000, 4.

52. R. Singh, "Patents amendment bill to protect traditional knowledge," *M2 Presswire*, 8 August 2000; "India to lead traditional knowledge database project," *BBC Worldwide Monitoring*, 11 July 2000.

53. P. Bereano, "Patent nonsense—Patent pending: The race to own DNA—Guaymi tribe was surprised to discover they were invented," *Seattle Times*, 27 August 1995: B5.

54. L. Andrews, observations at the First International Conference on DNA Sampling and Human Genetic Research: Ethical, Legal, and Policy Aspects: "DNA Sampling Conference," held at University of Montreal, Sept. 6–8, 1996.

55. B. O'Reilly, "Reaping a biotech blunder," *Fortune*, 19 February 2001, 156.

56. E. Pennisi, "Bracing for the war on cancer," *Science* 286 (1999): 24–31.

57. "The Reality of GMOs," *PCC Natural Markets*, available at http://www.pccnaturalmarkets.com/info/gmo.html.

58. J. Kluger et al., "Bad seeds: The battle heats up between the U.S. and Europe over genetically engineered crops," *Time*, 154 (20 September 1999): 12.

59. L. Andrews, supra note 3 at 261–262.

60. E. Goodman, "Genetic engineering, hustling collide," *Milwaukee Journal Sentinel,* 1 November 1999, 10.

61. L. Andrews, supra note 3 at 85.

62. Based on in vitro allergy results, Pioneer dropped the Brazil nut project. S. Schmickle, "Brazil nut project shows modified foods can inherit troubles of genes they receive," *Star Tribune,* Minneapolis, MN, 2 May 2000: 12A.

63. D.L.J. Freed, "Do dietary lectins cause disease?" *British Medical Journal* 318 (1999):1023–1024.

64. "The Reality of GMOs," supra note 57, citing report of the York Nutritional Lab, United Kingdom.

65. M. Lappé, B. Bailey, et al., "Alteration of clinically modified phytoestrogens," *Journal of Medicinal Foods* 1.4 (1999): 241.

66. M. Townsend, "Why soya is a hidden destroyer," *Daily Express,* East Malaysia, 12 March 1999.

67. "The Reality of GMOs," supra note 57.

68. R. Stone, "Large plots are next test for transgenic crop safety," *Science* 266 (1994): 1472. See also B.W. Faulk, "Will transgenic viruses generate new viruses and new diseases?" *Science* 263 (1994): 1423.

69. C.O. Tacket, H.S. Mason, G. Losonsky, M.K. Estes, M.M. Levine, and C.J. Arntzen, "Human immune responses to a novel Norwalk virus vaccine delivered in transgenic potatoes," *Journal of Infectious Diseases* 182 (2000): 302–305.

70. "Spuds that stop disease," *Wired News,* 12 July 2000, available at http://www.wired.com/news/print/0,1294,37525,00.html.

71. P. Bereano, "Body and soul: The price of biotech," *Seattle Times OpEd,* 20 August 1995, at B5, available available at: http://nativenet.uthscsa.edu/archive/nl/9511/0014.html.

Chapter Six. Taking Seriously the Claim That Genetic Engineering Could End Hunger: A Critical Analysis

1. See Council for Biotechnology Information, undated, *The promise of biotechnology: Food for a growing world population,* "Good Ideas Are Growing" Press Kit, at www.whybiotch.com ; M. McGloughlin, "Without biotechnology, we'll starve," *Los Angeles Times,* 1 November 1999, available at: http://www.biotech.ucdavis.edu/links/withoutbiotech.html; M. McGloughlin, "Ten reasons why biotechnology will be important to the developing world," *AgBioForum* 2.3, 2.4 (1996): 163–174, available at: http://www.agbioforum.org/vol2no34/mcgloughlin.html; P. Pinstrup-Andersen, "Biotech and the poor," *Washington Post,* 27 October 1999, A31.

2. F.M. Lappé, J.Collins, and P. Rosset with L. Esparza, *World Hunger: Twelve Myths,* Second Edition. (New York and London: Grove Press/Earthscan, 1998).

3. Ibid.

4. Ibid.

5. Ibid.

6. Ibid.

7. Ibid.

8. Ibid.

9. Ibid.; W. Bello, S. Walden, S. Cunningham, and B. Rau, *Dark Victory: The United States and Global Poverty*, Second Edition (London and Oakland: Pluto and Food First Books, 1999).

10. Lappé et al., *World Hunger*; Bello et al., *Dark Victory*.

11. Lappé et al., *World Hunger*.

12. Ibid.; also see debate between McGloughlin, 1999b, op. cit., and M.A. Altieri and P. Rosset, "Ten reasons why biotechnology will not ensure food security, protect the environment and reduce poverty in the developing world," *AgBio-Forum* 2.3, 2.4 (1999a): 155–162. On-line at: http://www.agbioforum.org/vol2no34/altieri.htm; and M.A. Altieri and P. Rosset, "Strengthening the case for why biotechnology will not help the developing world: response to McGloughlin," *AgBioForum* 2.3, 2.4 (1999b): 226–236. On-line at: http://www.agbioforum.org/vol2no34/altierireply.html

13. R.J.H. Chambers, "Farmer-First: A practical paradigm for the third world agriculture," in M.A. Altieri and S.B. Hecht, eds., *Agroecology and Small Farm Development* (Ann Arbor, MI: CRC Press, 1990), 237–244.

14. Ibid.

15. Chambers, "Farmer-First"; Lappé et al., *World Hunger*.

16. Chambers, "Farmer-First."

17. J. Jiggins, C. Reijnjets, and C. Lightfoot, "Mobilising science and technology to get agriculture moving in Africa: a response to Borlaug and Dowswell." *Development Policy Review* 14.1 (1996): 89–103.

18. Chambers, "Farmer-First."

19. Altieri and Rosset, 1999a; Altieri and Rosset, 1999b; ActionAid, *AstraZeneca and its Genetic Research: Feeding the World or Fueling Hunger?* (London: Action-Aid, 1999); Mae-Wan Ho, "The 'golden rice'—an exercise in how not to do science," *Third World Resurgence* 118/119 (2000): 22–26.

20. Altieri and Rosset, 1999a; Altieri and Rosset, 1999b.

21. Ibid.

22. Ibid.

23. Ibid.

24. Ibid.

25. Ibid.

26. Ibid.

27. Ibid.

28. M. Altieri, P. Rosset, and L.A. Thrupp, "The potential of agroecology to combat hunger in the developing world, Institute for Food and Development Policy," *Food First Policy Brief* 2 (1998); J. Pretty and R. Hine, "Reducing food poverty with sustainable agriculture: A summary of new evidence," *SAFE_World Project Final Report* (2001). On-line at: http://www2.essex.ac.uk/ces/ResearchProgrammes/ListofSusag.html.

29. Altieri et al., "The potential of agroecology."

30. P. Rosset, "The parable of the golden snail," *The Nation*, 27 December 1999; 22; I. Delforge, Nourrir le Monde ou L'Agrobusiness: Enquête sur Monsanto (Bruxelles: Les Magasins du Monde/Oxfam, 2000).

Chapter Seven. The European Response to GM Foods: Rethinking Food Governance

1. Gene Watch UK, GM Crops and Food: A Review of Developments in 1999. *Gene Watch UK Briefing, 9* (Tideswell: Gene Watch UK, 2000).

2. D. Robinson, H. Davies, A. Birch, T. Wilson, N. Kerby, G. Squire, and J. Hillman, "Development, release and regulation of GM crops," *Scottish Crop Research Institute Annual Report 1997/98* (Invergowrie: SCRI, 1998), 52.

3. Anon., "European Union okays use of soybean produced by biotechnology; case seen as turning point for industry," *World Food Regulation Review* 5(12) (1996): 27.

4. National Farmers Union, *Written Evidence of the NFU of England and Wales to the House of Lords' European Communities Committee Enquiry into EC Regulation of Genetic Modification in Agriculture* (London: National Farmers Union, 1998).

5. S. Nottingham, *Eat Your Genes: How Genetically Modified Food Is Entering Our Diet*, (London: Zed Books, 1998), 133.

6. *The Guardian Weekend*, 6 June 1998: 47.

7. Memorandum from J. Sainsbury plc to the House of Commons Select Committee on Agriculture, 15 November 1999; see http//www.publications.parliament.uk/pa/cm199900/cmselect/cmagric/71/7111.html.

8. Anon., "Boycott begins against products containing US genetically engineered soybeans," *World Food Regulation Review* 6(6) (1997): 22–23.

9. House of Commons Select Committee on Science and Technology (1999) Minutes of Evidence: Examination of Witnesses (Questions 1–19), 3 March 1999. See http://www.parliament.the-stationaryoffice.co.uk/pa/cm199899/cmselect/c.../9030303.html.

10. Anon., "Sainsbury's, M&S in 'GM-free' retailer consortium." *ENDS Report* 290 (1999): 33–34.

11. Friends of the Earth, "European food manufacturers shun GMOs but

consumers urged to keep up the pressure." Press Release 7 March 2000 (London: Friends of the Earth, UK).

12. House of Commons Select Committee on Agriculture, *The Segregation of Genetically Modified Foods. Third Report 1999–2000*. http://www.publications.parliament.uk/pa/cm199900/cmselect/cmagric/71/7103.html.

13. Rural Advancement Foundation International, "The Gene Giants: Masters of the Universe?" RAFI Communiqué (March/April 1999). Available at: http://www.rafi.org/communique/flxt/19992.html.

14. W. Heffernan, M. Hendrickson, and R. Gronski, *Consolidation in the Food and Agriculture System. Report prepared for the National Farmers Union* (Columbia: University of Missouri, 1999).

15. T. Marsden, A. Flynn, and M. Harrison, *Consuming Interests: The Social Provision of Foods* (London: UCL Press, 2000).

16. E. Marlier, *Eurobarometer* 35.1, "Opinions of Europeans on Biotechnology in 1991," in J. Durant, ed., *Biotechnology in Public: A Review of Recent Research* (London: Science Museum, 1992); E. Marlier, *Biotechnology and Genetic Engineering: What Europeans Think About It in 1993* (Brussels: INRA, 1993); European Commission, *The Europeans and Modern Biotechnology, Eurobarometer* 46.1, (Luxembourg: European Commission, 1997).

17. European Commission, *The Europeans and Modern Biotechnology, Eurobarometer* 46.1 (Luxembourg: European Commission, 1997).

18. S. Martin and J. Tait, "Attitudes of selected public groups in the UK to biotechnology," in J. Durant, ed., *Biotechnology in Public: A Review of Recent Research* (London: Science Museum, 1992).

19. A. Hamstra, *Consumer Acceptance of Food Biotechnology: The Relation Between Product Evaluation and Acceptance* (Den Haag: SWOKA, 1995).

20. L. Frewer, C. Howard, and R. Shepherd, "The influence of realistic product exposure on attitudes towards genetic engineering of food," *Food Quality and Preference* 7 (1996): 61–67; L. Frewer, C. Howard, and R. Shepherd, "Public concerns in the United Kingdom about general and specific applications of genetic engineering: Risk, benefit, and ethics," *Science, Technology and Human Values* 22 (1997): 98–124; L. Frewer, D. Hedderley, C. Howard, and R. Shepherd, "Objection mapping in determining group and individual concerns regarding genetic engineering," *Agriculture and Human Values* 14 (1997): 67–79; R. Grove-White, P. Mcnaughton, S. Mayer, and B. Wynne, *Uncertain World: Genetically Modified Organisms, Food and Public Attitudes in Britain* (Lancaster, UK: Centre for the Study of Environmental Change, Lancaster University, 1997).

21. R. Grove-White, P. Mcnaughton, S. Mayer, and B. Wynne, *Uncertain World: Genetically Modified Organisms, Food and Public Attitudes in Britain*

(Lancaster, UK: Centre for the Study of Environmental Change, Lancaster University, 1997).

22. A. Hamstra, *Consumer Acceptance of Food Biotechnology.*

23. Ibid.

24. R. Grove-White et al., *Uncertain World.*

25. L. Frewer et al., "Objection mapping"; A. Hamstra, *Consumer Acceptance of Food Biotechnology*; R. Grove-White et al., *Uncertain World.*

26. R. Grove-White et al., *Uncertain World.*

27. A. Hamstra, *Consumer Acceptance of Food Biotechnology.*

28. D. Barling, "GM crops, biodiversity and the European agri-environment: Regulatory regime lacunae and revision," *European Environment* 10 (2000): 167–177.

29. Friends of the Earth Europe, "EU Environment Council adopts 'de facto' moratorium on GMOs," *FoEE Biotech Mailout* 5.5 (1999): 7.

30. D. Barling, "GM crops, biodiversity and the European agri-environment."

31. Friends of the Earth Europe, "European Commission's initiative to end the GMO Moratorium," *FoEE Biotech Mailout* 6.5 (2000): 1.

32. *Official Journal of the European Communities* L. 43 (1997): 1–6.

33. M. Smith, "Novel foods create novel dilemmas for trade partners," *The Financial Times*, 22 June 1998: 7.

34. Anon., "Commissioner Bangemann concedes criticism of Novel Foods Regulation," *EU Food Law*, June 1997: 6.

35. Commission of the European Communities, *Proposal of the European Parliament and of the Council for a Regulation concerning the Traceability and Labelling of Genetically Modified Organisms and Traceability of Food and Feed Produced from Genetically Modified Organisms* (Brussels: European Commission, 2001). Commission of the European Communities, *Proposal for a Regulation of the European Parliament and of the Council on Genetically Modified Food and Feed,* (Brussels: European Commission, 2001).

36. Proposal for a regulation on Genetically Modified Food and Fee-Explanatory memo: 7.

37. D. Hencke and R. Evans, "How US put pressure on Blair over GM food," *Guardian*, 28 February (2000): 8.

38. N. Avery, M. Drake, and T. Lang, *Cracking the Codex: An analysis of Who Sets World Food Standards* (London: National Food Alliance, 1993).

39. T. Lang, D. Barling, and M. Caraher, "Food, Social Policy and the Environment: Towards a New Model," *Social Policy & Administration* 35 (2001): 538–558.

Chapter Eight. A Societal Role for Assessing the Safety of Bioengineered Foods

1. M. Lappè, "Reflections on the cost of doing science," *Annals of the Academy of Science* 265 (1976):45.

2. S. Davidson, "GM crops are no panacea for poverty," *Nature Biotechnology* 19 (2001): 797–799.

3. National Environmental Policy Act (NEPA), 42 USC 4331.

4. M. Lappé and P. Archbald, "The place of the public in the conduct of science," University of Southern California Law Review (1978) 51: 1539–1554.

5. SAP III, "Sets of scientific issues being considered by the Environmental Protection Agency regarding: Assessment of scientific information concerning Star-Link corn," FIFRA Scientific Advisory Panel, SAP Report No. 2000-06 (2000).

6. Food and Drug Administration, "Statement of Policy: Foods Derived from New Plant Varieties," FR57.104, 1992.

7. Alliance for Bio-Integrity, et al., V. Shalala, et al., No. Civ. A. 98 1300 (CKK). U.S. District Court, District of Columbia.

8. S. Druker, "U.S. court confirms that bioengineered foods are unregulated," 4 October 2000. Available at: http://www.biointegrity.org/PRESS_RELEASE-October4-2000.html.

9. G. Guest, Memo to James Maryanski, Biotechnology Coordinator, "Regulation of transgenic plants—FDA Draft Federal Register Notice on Feed Biotechnology," 5 February 1992.

10. S.R. Padgette, "The composition of glyphosate-tolerant soybean seeds is equivalent to that of conventional soybeans," *Journal of Nutrition* 126 (1999): 702–716.

11. R. Sidhu et al., "Glyphosate-tolerant corn: The composition and feeding value of grain from glyphosate-tolerant corn is equivalent to that of conventional corn (*Zea mays* L.)," *Journal of Agriculture and Food Chemistry* 48 (2000): 2305–2312.

12. B.G. Hammond et al., "The feeding value of soybeans fed to rats, chickens, catfish, and dairy cattle is not altered by genetic incorporation of glyphosate tolerance," *Journal of Nutrition* 126.3 (1996): 717–727.

13. B.G. Hammond et al., "The feeding value of soybeans fed to rats, chickens, catfish, and dairy cattle is not altered by genetic incorporation of glyphosate tolerance," *Journal of Nutrition* 126.3 (1996): 717–727.

14. Ibid.

15. R. Newbold, E. Banks, B. Bullock, W. Jefferson, "Uterine adenocarcinoma in mice treated neonatally with genistin," *Cancer Research* 61.11 (2001): 4325–4328.

16. M. Lappé, et al., "Alterations in clinically important phytoestrogens in geneti-

cally modified, herbicide-tolerant soybeans," *Journal of Medicinal Food* 1.4 (1999): 241–245.

17. N.B.Taylor, R.L. Fuchs, J. MacDonald, A.R. Shariff, and S.R. Padgette, "Compositional analysis of glyphosate-tolerant soybeans treated with glyphosate," *Journal of Agriculture and Food Chemistry* 47.10 (1999): 4469–4473.

18. U.S. EPA, "Biopesticides registration action document, *Bacillus thuringiensis* plant-incorporated protectants," Office of Pesticide Programs, 16 October 2001.

19. R. Goldberg, "Statement of the Environmental Defense Fund, the FDA public hearing on genetically engineered foods," 1999.

20. H.A. Sampson, "Food allergy: Immunopathogenesis and clinical disorders," *Journal of Allergy and Clinical Immunology* 103.5 (1999): 67–68.

21. EPA BRAD 2001, "Biopesticides registration action document: Revised risks and benefits sections—*Bacillus thuringiensis* plant pesticides," 16 July 2001.

22. SAP III, "Sets of scientific issues being considered."

Chapter Nine. Learning to Speak Ethics in Technological Debates

1. T. Berry, *The Great Work: Our Way Into the Future* (New York: Bell Tower, 1999), 24. See also B. Swimme and T. Berry, *The Universe Story* (San Francisco: Harper, 1992).

2. L. White Jr., "The historical roots of our ecological crisis," *Science* 155.3767 (1967): 1205.

3. D. Korten, "The Feasta Lecture," Dublin, Ireland, electronic version www.feasta.org/article_korten-civilizing.html 4 July 2000.

4. E.B. Tylor, *Primitive Culture* (New York: Harper & Row, 1958).

5. D.E. Hunter and P. Whitten, eds., *Encyclopedia of Anthropology* (New York: Harper & Row, 1976), 102–103.

6. R. Fulghum, *Everything I Really Need to Know I Learned in Kindergarten: Uncommon Thoughts on Common Things* (New York: Random House, 1988).

7. D. Korten, "Feasta Lecture."

8. P. Edwards, Ed., *Encyclopedia of Philosophy* (New York: McMillan Publishing Co., 1967), 81–82.

9. A. Bernstein, "Precaution and respect," in C. Raffensperger and J.Tickner, eds., *Protecting Public Health and the Environment: Implementing the Precautionary Principle* (Washington, DC: Island Press, 1999), 148.

10. Blue Mountain Participants: Andrew Jameton, Omaha, Nebraska; Bill Vitek, Potsdam, New York; Bruce McKay, Montreal, Quebec; Carolyn Raffensperger, Windsor, North Dakota; Craig Holdrege, Ghent, New York; David Abram, Victor, Idaho; Derrick Jensen, Crescent City, California; Fred Kirschenmann, Ames, Iowa; Harriet Barlow, Minneapolis, Minnesota; Jennifer Sahn, Great

Barrington, Massachusetts; Katherine Barrett, Victoria, British Columbia; Maria Pellerano, Annapolis, Maryland; Marianne Spitzform, Missoula, Montana; Mary O'Brien, Eugene, Oregon; Mark Ritchie, Minneapolis, Minnesota; Nancy J. Myers, Oak Park, Illinois; Peter deFur, Richmond, Virginia; Peter Montague, Annapolis, Maryland; Peter Sauer, Salem, New York; Sheila Kinney, Blue Mountain Lake, New York; Steve Light, Minneapolis, Minnesota; Ted Schettler, Boston, Massachusetts; Tracey Easthope, Ann Arbor, Michigan; Wes Jackson, Salina, Kansas.

Chapter Ten. A Perspective on Anti-Biotechnology Convictions

1. The Food and Drug Administration does not have a regulatory definition of "substantial equivalency": instead, it leaves it up to the producer or manufacturer to indicate whether or not a new product differs substantially or contains an additive that makes it nonequivalent to a previous product.

2. K. Eichenwald, G. Kolata, and M. Petersen, "Biotechnology food: From the lab to a debacle," *New York Times* http://www.nytimes.com/2001/01/25/business25FOOD.html.

3. Research conducted by S. R. Padgette and his colleagues was sent to the FDA in 1994 and put into print a year and a half later; see S.R. Padgette, N.B. Taylor, D.L. Nida, M.R. Bailey, J. MacDonald, L.R. Holden, and R.L. Fuchs, "The composition of glyphosate-tolerant soybean seeds is equivalent to that of conventional soybeans," *Journal of Nutrition* 126 (1996): 702–716.

4. M.A. Lappé, E.B. Bailey, C. Childress, and K.D.R. Setchell, "Alterations in clinically important phytoestrogens in genetically modified, herbicide-tolerant soybeans," *Journal of Medicinal Food* 4 (1999): 241–246.

5. See original submission to the *Federal Register,* 6 December 1992, 58 (232): Notice 93-148-1.

6. 90/220/EEC and 90/219/EEC, respectively.

7. 49/2000/EEC and 50/2000/EEC, respectively.

8. K. Eichenwald et al., "Biotechnology food."

9. Ibid.

10. See http://truefood.org. 17 January 2001.

11. National Research Council, *Genetically Modified Pest-Protected Plants* (Washington, DC: National Academy Press, 2000), 2.

12. *Chemical Market Reporter* for 18 June 2001.

13. See www.nytimes.com/2000/12/14/science/14BIOT.html.

14. See *Mite Fax: San Joaquin Valley Cotton* (Brandon, MS: Looking South Communications, 30 June 2001).

15. S.G. Rogers, "Biotechnology and the soybean," *American Journal of Clinical Nutrition* 68 (1998): 1330S–1332S.

16. See http://hoovenews.hoovers.com/fp/asp?layout=displaynews&doc_id=NR 20010619670.2_75b7000840ab7e3d.html; 19 June 2001.

17. See http://dailynews.yahoo.com/h/ap/20010117/pl/biotech_food_1.html.

18. R. Lewontin, "Genes in the food!" *New York Review of Books* XLVIII (2001) 10: 81–85.

19. See "NRC acknowledges GM food risks," *Science* 11 December 2000.

20. M. Lappé and B. Bailey, *Against the Grain: Biotechnology and the Corporate Takeover of Your Food* (Monroe, ME: Common Courage Press, 1998).

21. "Cost worries block accord on GM food traceability." www.organicTS.com news, 3 April 2001.

22. See www.CropChoice.com 18 June 2001.

23. See http://ens-news.com/ens/jun2001/2001L-06-18-09.html.

24. J. Schnittker, quoted by C. Abbott, "U.S. farmers tightening embrace of GM crops," http://www.planetark.org/dailynewsstory.cfm?newsid=10332; 2 April 2001.

25. "U.S. withdraws genetically engineered corn-animal feed donation after Bosnia's hesitation," available at: http://www.centraleurope.com/bosniatodaynews.php3?id=273802; 30 January 2001.

26. See "Special Issue: Food Ethics and Consumer Concerns," *Journal of Agricultural and Environmental Ethics* 2 (2000): 12.

27. R. Newbold, E. Banks, B. Bullock, W. Jefferson, "Uterine Adenocarcinoma in Mice Treated Neonatally with Genistin," *Cancer Research* 61.11 (2001): 4325-4328.

28. Published at www.nytimes.com/2000/12/14/science/14BIOT.html.

29. Minutes of the USDA, Advisory Committee Meeting, August 1–2, 2001, file://C:\WINDOWS\TEMP\ppbprpt_8-01.html.

30. R. Lewontin, "Genes in the food!" *New York Review of Books* XLVIII (2001) 10: 84.

Appendix A. In Defense of the Precautionary Principle

1. http://www.foodsafety.gov/~fsg/fssyst4.html.

2. http://europa.eu.int/comm/external_relations/us/biotech/report.pdf

3. http://www.rsc.ca/foodbiotechnology/indexEN.html.

About the Contributors

LORI ANDREWS is professor of law at Chicago-Kent College of Law, and director of the Institute for Science, Law, and Technology at Illinois Institute of Technology. In her policy work on genetic technologies, she has served on advisory committees to the World Health Organization, the National Institutes of Health, the Centers for Disease Control, the U.S. Department of Health and Human Services, and the National Academy of Sciences. She has served and chaired the federal Working Group on the Ethical, Legal and Social Implications of the Human Genome Project. Andrews is the author of seven books and more than a hundred articles on genetics, alternative modes of reproduction, and biotechnology. She was named one of the most influential lawyers in America by the *National Law Journal*.

BRITT BAILEY is the senior associate at the Center for Ethics and Toxics in Gualala, California. In that capacity, she has researched and written a number of key articles on the misuse of pesticides, mad cow disease, transgenic crops, the history of agriculture, and the ethical issues surrounding biotechnology. She co-authored *Against the Grain: Biotechnology and the Corporate Takeover of Your Food* (1998) and is the producer of *Against the Grain: The Video* (1999). She has testified before the California State Senate Agriculture and Trade Committee advocating the position of the public's right to know they are eating foods containing genetically engineered components. She holds a master's degree in environmental policy and is also an instructor in the Life Sciences Department at the College of Marin in Kentfield, California.

DAVID BARLING is a senior lecturer at the Center for Food Policy at Thames Valley University, London, where he is the program leader for the MA in Food Policy. His research focuses on the governance of food and agriculture, notably the regulation of genetic modification throughout the food chain. He has published a number of book chapters and reports as well as articles in journals such as *European Journal of Public Health, European Environment, Environmental Toxicology and Pharmacology,* and *Social Policy & Administration.*

NORMAN C. ELLSTRAND is a professor of genetics at the University of California at Riverside. At age four, he wowed his parents by unexpectedly identifying pronghorn antelopes during a South Dakota vacation. His primary teaching effort has been "Human Heredity for Nonmajors," but his research theme is plant population genetics. His current research emphasis is on the consequences of gene flow from domesticated plants to their wild relatives, including the escape of engineered genes. Although Ellstrand secretly wants to write a novel, he has had fun being involved with science. He has written over a hundred papers, has talked to congressional staff about his research, participated in a number of activities of the National Research Council (including a study of environmental impacts associated with the commercialization of transgenic plants), and lived in Sweden for four months on a Fulbright Fellowship. In his spare time, he helps create new units on his campus, such as the Center for Conservation Biology and the proposed Biotechnology Impacts Center. He is married to Dr. Tracy Kahn, curator of University of California at Riverside's Citrus Variety Collection; they have one son, Nathan, who, at age fourteen, is considering both law and theme park design as possible career choices.

BREWSTER KNEEN is the author of five books and publisher of *The Ram's Horn,* a monthly newsletter of food system analysis. He was born in Ohio, and studied economics and theology in the U.S. and the U.K. before moving to Toronto in 1965. There, he produced public affairs programs for CBC Radio, renovated houses, and worked as a consultant to the churches on issues of social and economic justice. Kneen has been writing and lecturing on the food system, with increasing attention to biotechnology. He was a founding member of the Toronto Food Policy Council and served on it from 1990 to 1993. In 1994–1995, he was a Senior Fellow of the Faculty of Environmental Studies, York University. Kneen's latest book, *Farmageddon: Food and the Culture of Biotechnology,* was published by New Society Publishers in 1999. A French edition, *Les aliments trafiqués,* was published by Ecosociété in

Montreal in June 2000. The Kneens continue to publish *The Ram's Horn* for an ever-growing global readership.

SHELDON KRIMSKY is professor of urban and environmental policy at Tufts University. He received his bachelor's and master's degrees in physics from Brooklyn College, CUNY, and Purdue University, respectively, and a master's and doctorate in philosophy from Boston University. Professor Krimsky's research has focused on the linkages between science/technology, ethics/values, and public policy. He served on the National Institutes of Health's Recombinant DNA Advisory Committee from 1978 to 1981. He was a consultant to the Presidential Commission for the Study of Ethical Problems in Medicine and Biomedical and Behavioral Research, and to the Congressional Office of Technology Assessment. Currently, he serves on the board of directors for the Council for Responsible Genetics and as a Fellow of the Hastings Center on Bioethics.

FRANCES MOORE LAPPÉ is the author of twelve books, including the international best-seller *Diet for a Small Planet*. Her books have been used in countless university courses and translated into twenty-two languages. In 1987, she became the fourth American to receive the Right Livelihood Award, often called the "Alternative Nobel." She is also a co-founder of the twenty-six-year-old Institute for Food and Development Policy/Food First, based in Oakland, California, as well as the Center for Living Democracy, a ten-year initiative to further citizen problem solving in every dimension of public life. Her forthcoming book, co-authored with her daughter Anna, is *Hope's Edge: The Next Diet for a Small Planet*. It will include recipes from many pioneer American vegetarian cookbook authors and leading vegetarian chefs. www.dietforasmallplanet.com.

MARC LAPPÉ directs the Center for Ethics and Toxics in Gualala, California. He has been professor of health policy and ethics at the University of Illinois College of Medicine. He also headed the Hastings Center's Genetic Research Group (1971–1976), which published a set of principles at stake in genetic screening. He has published or edited 14 books in genetics, occupational and public health as well as more fundamental issues relating to scientific research. His current work centers on ag-biotech and ethics, preventing toxic exposures, and health policy in general. He received his bachelor's degree in biology from Wesleyan University and his doctorate in experimental pathology from the University of Pennsylvania.

CAROLYN RAFFENSPERGER is the executive director of the Science and Environmental Health Network, a consortium of environmental organizations dedicated to using science to protect public health and the environment. Raffensperger holds a master's in archaeology and a law degree. Her specialty is in the philosophy of science and its practical applications to the problems facing society today. She is the co-editor of *Protecting Public Health and the Environment: Implementing the Precautionary Principle* (Island Press, 1999). This volume provides the history and methods for applying the precautionary principle. Raffensperger lives on a 3,500-acre farm in North Dakota.

PETER ROSSET is co-director of Food First/The Institute for Food and Development Policy, based in Oakland, California (www.foodfirst.org). He has a Ph.D. in agricultural ecology from the University of Michigan, and is an internationally recognized expert on agricultural technology and development. Among his many books are *Agroecology* (McGraw-Hill, 1990) and *World Hunger: Twelve Myths*, Second Edition (Grove Press and Earthscan, 1998).

PAUL THOMPSON holds the Joyce and Edward E. Brewer Chair in Applied Ethics at Purdue University, where he is also a Distinguished Professor of Philosophy. He received his Ph.D. in philosophy from the State University of New York at Stony Brook in 1980, and served for sixteen years on the faculties in philosophy and agricultural economics at Texas A&M University. He is the author or editor of seven books, including *The Agrarian Roots of Pragmatism* (Vanderbilt University Press, 2000), *Food Biotechnology in Ethical Perspective* (Chapman and Hall, 1997), and *The Spirit of the Soil: Agriculture and Environmental Ethics* (Routledge Publishing Co., 1995). He has published more than a hundred papers on ethical issues associated with agriculture, risk, and genetic technology.

Index

Aboriginal communities, 51
Abortion, 71, 174n18
Accountability, scientific: consumer right to, 6, 148–49; corporate, 5, 7, 116–17, 119
Acreage of GM crops, 6, 12, 96–97, 119, 120
Activist community. *See* Environmental activists
Africa, 54–55; sub-Saharan, 82, 86–87, 88
Against the Grain (Bailey and Lappé), 6, 140, 148
Agaricus bisporus, 64
Agency: European consumers and, 101, 102; progress and, 52, 55–59
Agricultural biotechnology. *See* Biotechnology industry
Akre, Jane, 140
Alberta, Canada, 65
Alleles, 62–65
Allergenicity: battle lines drawn over, 5–6, 9; in Bt crops, 142–44; in GM soybeans, 118, 176n62
Allergen transfer: FDA labeling policies for, 17, 141; food safety and, 38–39; in GM soybeans, 14, 77–78; right to exit and, 40; in StarLink corn, 121–22, 148–49
Alliance for Biointegrity, 116–17
"All I Really Need to Know I Learned in Kindergarten" (Fulghum), 128
American Association for the Advancement of Science, 153
American Soybean Association, 98, 119
Amish, 51, 170n4
Ampicillin, 96–97

Andrews, Lori, 3
Animal feed: in Europe, 106, 107; in food safety tests, 118; StarLink corn approved as, 3, 121–22, 148–49
Anti-biotechnology movement, 2, 39–40, 95–111, 135–56. *See also* Environmental activists; NGOs
Antibiotic-resistant genes: ethical issues concerning, 11, 22, 68; in Europe, 96–97, 104; FDA labeling policies for, 141; food safety and, 151; in Great Britain, 147
Antibiotics: overuse of, 22; in sterile seed technology, 20
Antwerp (Belgium), 97
APHIS program, 3
Approval of GMOs: in Europe, 103–7; by regulatory agencies, 3–4, 96, 171n11; regulatory philosophies of biotech industry and, 3, 51–52, 56
Archer Daniels Midland (ADM), 99, 108
Argentina, 96
Aristotelian principle, 136
Artists, 130
Asgrow, 69
Ashkenazi Jewish community, 73, 175n38
Asian pear, 2, 150–51
Astra Zeneca, 99
Athletes, professional, 74, 175–76n44
Aunt Orva, 27–31, 33–34, 38, 39, 42–43
Australia, 4, 16, 161
Australian rye grass, 153
Austria, 105
Aventis CropSciences, 65, 99, 121, 144, 149, 171n11